Virtue Ethics and the Environment

Philip Cafaro
Ronald Sandler
Editors

Virtue Ethics and the Environment

Previously published in the *Journal of Agricultural and Environmental Ethics*, Volume 23(1–2), 2010

Springer

Editors
Philip Cafaro
Colorado State University
Fort Collins, USA

Ronald Sandler
Northeastern University
Boston, USA

ISBN: 978-94-007-0226-4
Springer Dordrecht Heidelberg London New York

Library of Congress Control Number: 2010936505

Printed on acid-free paper

Springer.com

Contents

RICHARD P. HAYNES / Editorial 1

PHILIP CAFARO / Environmental Virtue Ethics Special Issue: Introduction 3–7

BRIAN TREANOR / Environmentalism and Public Virtue 9–28

KATHRYN J. NORLOCK / Forgivingness, Pessimism, and Environmental Citizenship 29–42

ALLEN THOMPSON / Radical Hope for Living Well in a Warmer World 43–59

JEREMY BENDIK-KEYMER / Species Extinction and the Vice of Thoughtlessness: The Importance of Spiritual Exercises for Learning Virtue 61–83

JOSHUA COLT GAMBREL and PHILIP CAFARO / The Virtue of Simplicity 85–108

JASON KAWALL / The Epistemic Demands of Environmental Virtue 109–128

PAUL HAUGHT / Hume's Knave and Nonanthropocentric Virtues 129–143

CHRISTINE SWANTON / Heideggerian Environmental Virtue Ethics 145–166

RONALD SANDLER / Ethical Theory and the Problem of Inconsequentialism: Why Environmental Ethicists Should be Virtue-Oriented Ethicists 167–183

PHILIP CAFARO / Patriotism as an Environmental Virtue 185–206

J Agric Environ Ethics (2010) 23:1
DOI 10.1007/s10806-009-9206-1

Editorial

Richard P. Haynes

Published online: 19 August 2009

From the editor

This special issue of the *Journal of Agricultural and Environmental Ethics* is dedicated to Dr. Marilyn Holly, a colleague of mine in the University of Florida Philosophy Department for 40 years. I owe my interest in environmental virtue ethics to her discussions with me in preparation for a review essay she was writing on this topic. The essay, published in volume 19, no. 4 (2006), discussed Lisa Newton's *Ethics and Sustainability: Sustainable Development and the Moral Life;* Philip Cafaro's *Thoreau's Living Ethics: Walden and the Pursuit of Virtue,* and Ronald Sandler and Philip Cafaro's collection, *Environmental Virtue Ethics.* Dr. Holly and I were jointly teaching a course in feminist philosophy in the summer of 2007 when she passed away on July 31 (exactly 2 years ago). I am writing this editorial on the second anniversary of her death. After she died, I acquired her copy of the Sandler-Cafaro collection, and after going through it decided to ask its editors if they would be willing to put together a sequel as a special issue of this journal. They agreed, and I started receiving drafts of the papers of the invited contributors in December of 2008. I put these papers through the normal review process and received all of the final versions by July of this year. Cafaro agreed to write an introductory piece, which is the first article in this collection. I am grateful to both Phil and Ron for the work they put into this collection and also to the various contributors of these essays. I hope that the reader will find them as interesting and thought provoking as I have.

References

Holly, M. (2006). Review essay: Environmental virtue ethics, a review of some current work. *Journal of Agricultural and Environmental Ethics, 19*(4), 391–424.

R. P. Haynes (✉)
Agriculture, Food, and Human Values Society, Gainesville, FL 32608, USA
e-mail: rhaynes@phil.ufl.edu

J Agric Environ Ethics (2010) 23:3–7
DOI 10.1007/s10806-009-9204-3

ARTICLES

Environmental Virtue Ethics Special Issue: Introduction

Philip Cafaro

Accepted: 27 May 2009 / Published online: 11 August 2009

Environmental virtue ethics (EVE) first got off the ground in 1983, with Thomas Hill, Jr.'s "Ideals of human excellence and preserving natural environments." Hill suggested that rather than trying to prove nature's intrinsic value or justify nonhuman rights, we might develop an environmental ethics by asking, in the face of environmentally destructive behavior: what kind of a person would do that? And: what kinds of persons will we become if we do likewise?

Or perhaps EVE really started in 1853, with publication of Henry Thoreau's (1971) *Walden*, which lays out an approach to life focused on "plain living and high thinking" and connection to nature. It explicitly contrasts such a life with the materialism that has subsequently gotten us into so much environmental trouble. "When [a person] has obtained those things which are necessary to life," Thoreau reminds us, "there is another alternative than to obtain the superfluities; and that is, to adventure on life now, his vacation from humbler toil having commenced."

Unless, of course, EVE started in the mid-Fourth century BC, with the composition of Aristotle's *Politics*: "Since one should take thought for the health of the [city's] inhabitants, this consists in the location being finely situated…and second, in using healthy sorts of water, and making this more than an incidental concern; for the things we use most of and most often for the body are what contribute most to health, and the capacity of waters and wind has such a nature" (book 7, chapter 11). Or perhaps it began a century or so earlier, half a world away, in the sayings eventually collected in the *Tao Te Ching*, with their emphasis on non-doing (*wu wei*) and acceptance of the world as it is:

> Oh, for a small country with few people!
> …Although there exist boats and carriages, they have no occasion to ride in them.

P. Cafaro (✉)
Philosophy Department, Colorado State University, Fort Collins, CO 80523, USA
e-mail: philip.cafaro@colostate.edu

> Although there exist armor and weapons, they have no occasion to show them off.
> ...They find their simple food savory and their simple clothes beautiful,
> They are content with their homes and happy in the way they live. (book 2, chapter 80).

Whatever its origins, EVE now appears to be a vital part of the environmental philosophy landscape. Whether you look to it for an alternative to other theoretical approaches to environmental ethics, or as a complement to them, there are a number of ways EVE clarifies our environmental choices and responsibilities:

- By deploying a multifaceted set of virtue and vice concepts, it provides a rich, nuanced descriptive and normative language for our relationships and interactions with the natural environment (as described in chapter 1 of Louke van Wensveen's 2000 *Dirty virtues*).
- By explicating the connection between human flourishing and nature, it complements duty-based and fear-based justifications for environmentally progressive behavior and policies.
- By providing an ecologically informed account of human flourishing that is attentive to the full range of environmental values, it offers an alternative to consumption-oriented conceptions of human flourishing.
- By articulating a positive, aspirational vision in which humans and nature flourish together, it provides an alternative conception of the human-nature relationship to those in which people are either villains and despoilers, or self-denying ascetics.
- By specifying the kind of character conducive to environmental appreciation and personal restraint, it contributes to sketching the parameters of a genuinely sustainable society: one which doesn't careen from one environmental problem to the next, seeking techno-fixes, but instead might really solve them.
- By focusing attention on the character states necessary for accomplishing lasting environmental improvement, it raises the salience of moral development and education to environmental ethics.

In these ways, EVE has proven its worth. With the recent publication of Ronald Sandler's (2007) *Character and environment*, the field has its most theoretically rigorous formulation to date. It is a good foundation on which to build, as several contributions to the present collection illustrate.

This special issue leads off with Brian Treanor's "Environmentalism and public virtue." Treanor notes that EVE has tended to focus on issues of personal virtue and individual benefit. "There has, in contrast, been relatively little interest in either virtue politics—collective actions, characteristics, or dispositions—or in public virtues: actions, characteristics, or dispositions that benefit the community rather than the individual." This seems a fair criticism; his piece begins to remedy this deficiency in EVE.

Kate Norlock, in "Forgivingness, pessimism and environmental citizenship," asks, given the great crimes against nature and future generations being committed today, how we can keep from falling into despair and pointless, teeth-grating anger.

Norlock finds forgiveness a necessary environmental virtue, one that helps us respond to ongoing ecological destruction in a productive way. She also asks when "rational pessimism" and its attendant political quietism might actually be reasonable.

Allen Thompson's "Radical hope for living well in a warmer world" explores the virtues conducive to getting through the "radical cultural changes" likely to be precipitated by global climate change. Thompson argues that persevering in the face of massive environmental degradation will demand great intellectual creativity and practical flexibility, as well as the radical hope that positive cultural forms will be found on the other side.

Jeremy Bendik-Keymer's "Species extinction and the vice of thoughtlessness" takes its departure from the Aristotelian reminder that virtue must be habituated. Modeled on a form of spiritual exercise used by the ancients, it grapples with the current extinction crisis in a way that does justice to the gravity of the issue and seeks to foster an active response to it. Bendik-Keymer's approach reminds us that an essential part of environmental philosophy is making the move from factual knowledge and ethical judgment to action. He also raises challenging questions about the role of theory in environmental ethics and moral philosophy more broadly.

Environmental degradation is primarily a matter of too many people consuming too much energy and materials. Joshua Gambrel's and Philip Cafaro's "The virtue of simplicity" argues that we can live better lives while (and in many cases *by*) consuming less. Material simplicity is a personal virtue, useful for promoting individual flourishing. In a crowded, overdeveloped world, simplicity is also a necessary virtue if we hope to do right by nature and fairly share limited resources with the world's poor.

Jason Kawall's "The epistemic demands of environmental virtue" addresses a crucial issue at the intersection of ethics and epistemology, particularized to EVE: how to act virtuously under conditions of uncertainty. Living an environmentally virtuous life requires information—about ethics and ecology, for example—but we are finite beings with limited time and resources, who often cannot act with as much information as we might like. Kawall explicates the epistemic demands of environmental virtue and provides rules of thumb regarding what issues we should prioritize in gathering information, what sort of information we ought to seek, and what level of information we ought to accept.

In "Hume's knave and nonanthropocentric virtues," Paul Haught challenges contemporary nonanthropocentric EVE with a critique developed from Hume's discussion of the sensible knave. Haught argues that, like other forms of nonanthropocentric environmental ethics, nonanthropocentric EVE may find it difficult to engage those whose virtues incline them to anthropocentrism. Nonetheless, he believes that EVE is perhaps better off than other nonanthropocentric environmental ethics, since it may be able to bridge the motivational gap between anthropocentrism and nonanthropocentrism through the explication of specific virtues.

In *Virtue ethics: a pluralistic view*, Christine Swanton (2003) suggested that appreciation of nature and respect for nonhuman beings were important virtues. However, taking this suggestion seriously leads to an apparent dilemma. On the one hand, valuing nature for itself seems necessary, in order to do justice to our environmental responsibilities; on the other hand, the demands of living in the world seem to push us in a more anthropocentric direction. In "Heideggerian environmental virtue ethics," Swanton argues that Martin Heidegger's philosophy provides one plausible means of working through this tension and dwelling respectfully on Earth.

In "Ethical theory and the problem of inconsequentialism: why environmental ethicists should be virtue-oriented ethicists," Ronald Sandler claims that many significant environmental problems are longitudinal collective action problems. Given that an individual's contribution is nearly inconsequential to addressing an environmental problem and may require some cost from the standpoint of her own life, why should she make the effort—particularly when it is uncertain (or even unlikely) that others will do so? Sandler argues that an adequate environmental ethics must respond cogently to this problem and that virtue-oriented ethical theory supports several plausible responses, whereas non-virtue-oriented forms of utilitarianism and Kantianism do not.

Finally, in "Patriotism as an environmental virtue," Philip Cafaro waves the flag for a local, place-based patriotism, as a key to environmental appreciation and protection. Knowing and caring about particular places and people and working to protect them is good for us and good for them and hence a good thing overall. Knowing and caring and working less on behalf of more remote places and people is also good, since it allows us to focus our efforts, act effectively, and do more good in the world. One benefit of developing a green patriotism is its potential for uniting liberals and conservatives around environmental protection.

We hope you will agree that this is an exciting collection that addresses compelling issues, not only in EVE, but in environmental philosophy more broadly. In addition to their individual merits, we believe these articles collectively demonstrate the value of EVE to environmental philosophy and the value of environmental philosophy in helping create sustainable societies and preserve nature. More work remains to be done, however, if EVE is to fulfill its full promise.

Echoing Treanor, we believe that high on the agenda should be more robust engagement with political issues from an EVE perspective. Further specifying modes of human flourishing that are truly sustainable and just to the ten million other species with whom we grudgingly share our planet is also key. Recent movement on negotiating a strong successor to the Kyoto Treaty gives some hope that the world is waking up to the moral imperative to protect the Earth. But whether people will be willing to ratchet down our economic demands and reshape our economies to reflect this hoped-for moral progress remains to be seen.

References

Hill, T. (1983). Ideals of human excellences and preserving natural environments. *Environmental Ethics, 5*(3), 211–224.

Sandler, R. (2007). *Character and environment: a virtue oriented approach to environmental ethics*. New York: Columbia University Press.
Swanton, C. (2003). *Virtue ethics: a pluralistic view*. Oxford: Oxford University Press.
Thoreau, H. D. (1971). *Walden*. Princeton: Princeton University Press.
Wensveen, L. V. (2000). *Dirty virtues: the emergence of ecological virtue ethics*. Amherst: Prometheus.

J Agric Environ Ethics (2010) 23:9–28
DOI 10.1007/s10806-009-9184-3

ARTICLES

Environmentalism and Public Virtue

Brian Treanor

Accepted: 27 May 2009 / Published online: 16 June 2009

Abstract Much of the literature addressing environmental virtue tends to focus on what might be called "personal virtue"—individual actions, characteristics, or dispositions that benefit the individual actor. There has, in contrast, been relatively little interest in either "virtue politics"—collective actions, characteristics, or dispositions—or in what might be called "public virtues," actions, characteristics, or dispositions that benefit the community rather than the individual. This focus, however, is problematic, especially in a society that valorizes individuality. This paper examines public virtue and its role in environmental virtue ethics. First, I outline different types of virtue in order to frame the discussion of public virtues and, in particular, a subclass of virtues I will refer to as political virtue. Second, I focus on practical problems and address the inadequacy of personal virtue for effecting social change and, therefore, for addressing most environmental crises. Finally, I argue that public and political virtues are necessary, if under emphasized, conditions for the flourishing of the individual, and that they are important complements to more traditional environmental virtues.

Keywords Environmental virtue · Civic virtue · Politics · Community · The tragedy of the commons · Flourishing

A community is like a ship; everyone ought to be prepared to take the helm
—Henrik Ibsen

B. Treanor (✉)
Department of Philosophy, Loyola Marymount University, One LMU Drive, Los Angeles, CA 90045, USA
e-mail: btreanor@lmu.edu

Introduction

Much of the literature addressing environmental virtue tends to focus on what might be called "personal virtue"—individual actions, characteristics, or dispositions that benefit the individual actor.[1] There has, in contrast, been relatively little interest in either "virtue politics"—collective actions, characteristics, or dispositions—or in what might be called "public virtues," individual actions, characteristics, or dispositions that benefit the community rather than the individual. This, however, is unfortunate. Environmentalists seem to take for granted that virtues such as simplicity, respect, or care might function as cardinal environmental virtues (Treanor 2008; Wensveen 2000); however, upon consideration, there is good reason to think that public virtues—especially the subset of political virtues—play the role of a keystone in environmental virtue, without which the whole edifice of environmental virtue is doomed to fall.[2]

This emphasis on personal virtue has not gone unnoticed. Garrett Hardin asserted that, except for some "valuable comments" made by Hans Jonas in his essay "Technology and Responsibility," "ethical literature is almost wholly individualistic: it is addressed to private conduct rather than public policy" (Hardin 2008, p. 350).[3] The tendency to ignore virtue politics is, perhaps, understandable. As Phil Cafaro notes, "it is easier to see how one individual can act on a noble vision of the good life than how a whole citizenry might" (Cafaro 2004, p. 195).[4] Public virtue has fared slightly better than virtue politics. Aristotle addressed several public virtues, including truthfulness, friendship, and justice; and several contemporary ethicists also deal with public virtues.[5] However, while all environmentalists are concerned with real-world environmental problems and at least some environmental virtue ethicists acknowledge the necessity of public virtue broadly conceived, the

[1] My thanks to Dr. Gitty Amini and Ms. Whitney Chelgren, as well as Phil Cafaro, Ronald Sandler, and the anonymous reviewers for *The Journal of Agricultural and Environmental Ethics*, for their helpful comments on an earlier draft of this paper.

[2] "Political virtues" are distinct from "virtue politics," as will become clear in what follows. In brief, political virtues are virtues possessed by an individual that express themselves in the political arena, broadly defined. Virtue politics, in contrast, would be the attempt to think of public policy and group action in terms of virtue and vice. Sofia Vaz—of the Center for Environmental and Sustainability Research at the New University of Lisbon, Portugal—offered an intriguing analysis of the possibility of a virtue politics in her "Conditions of Possibility of a Virtue Politics Applied to Responsible Consumption," a paper presented at the 2008 meeting of the International Association for Environmental Philosophy (IAEP).

[3] While it is true that Hardin made this comment in 1977 and that there has been increased interest in environmental public policy since that time, it remains the case that public conduct and public virtue is under-analyzed in environmental virtue ethics.

[4] Thoreau himself only "imagines such societies fleetingly and poetically" (Cafaro 2004, p. 196).

[5] Among thinkers who do address public virtues we might mention Aristotle (*The Nicomachean Ethics, Politics*), Sandler (*Character and Environment*), Dagger (*Civic Virtues*), and others. In addition, one of the reviewers for the JAEE brought to my attention James Connelly's forthcoming *Sustainability and the Virtues of Environmental Citizenship* (Routledge 2009), which appears to address the matter at issue here. However, while these and other works do address public and political virtue, I'm still struck by the *emphasis* on personal virtues exhibited in both academic and, especially, popular discourse, as Sandler himself points out in the next citation.

emphasis remains squarely on personal virtue. Even among thinkers who do address *public* virtue, few address the more specific need for *political* virtues such as political engagement and activism.

> Virtues of environmental activism have been somewhat overlooked and underappreciated by environmental ethicists in comparison with other environmental virtues. For those of us raised on North American environmentalism in particular, it is tempting to contemplate our environmental heroes as walkers in the woods or hikers of the mountains. The image is of an individual with personal excellence flourishing in harmony with and appreciation of the wild (or near wild) environments (Sandler 2007, p. 49).[6]

This disconnect between our heroes and exemplars and our current social, political, and environmental reality has some unfortunate consequences, leading Louke Van Wensveen to characterize environmental virtue ethics as a "visionary discourse" that "bears witness to a social vision," but which is, ironically, "without a social ethic" (Wensveen 2005, p. 26).

Among academics the tendency to think of virtue in terms of the individual is exacerbated by the predominance of articulating environmental virtues within a eudaimonistic framework.[7] Although there are social components to a eudaimonistic view of virtue, to which I will return below, these are generally articulated in terms of the flourishing of the individual. For example, Aristotle says that while the essence of friendship is loving rather than being loved, it is nevertheless the case that friendship is based on the love of oneself.[8] His eudaimonistic approach to virtue tends to focus on individual conduct and the effect that cultivating virtue has on the individual. On this account virtues are the means to the end of personal flourishing. This is not to imply that virtues are *not* eudaimonistic; they certainly are. However, a eudaimonistic framework shapes the discourse about virtue in a particular way that tends to deemphasize the disinterested aspects of *eudaimonia*.

Outside academic discourse there is an even stronger tendency to think of environmental virtue in terms of an optional and personal—rather than necessary and socially engaged—choice. Vice President Dick Cheney famously claimed that environmentalism is merely a "personal virtue,"[9] implying such dispositions are something that might contribute to the personal happiness of those so inclined, precisely and only because they are so inclined, but which are neither essential for flourishing nor social in scope. The suggestion seems to be that environmental virtues are something like upper class social graces or exceptional penmanship—valuable to those for whom they matter, but only to those for whom they matter.

[6] Sandler treats the social context of flourishing and/or the role of activism in several passages (e.g., Sandler 2007 , pp. 45–46 and 49–50). His "pluralistic teleological account" of the virtues makes particular mention of both the good functioning of the social group and of noneudaimonistic ends (Ibid., 28, 63–83).

[7] This is an observation rather than a criticism, as my own approach to virtue ethics tends to be eudaimonistic in character.

[8] For example, the good man "wishes for himself what is good" (Aristotle 1925, p. 227) and is a "lover of self" (Aristotle 1925, p. 237).

[9] Dick Cheney made this claim in a speech in Toronto, Canada on April 30, 2001.

We might dismiss Cheney's comment as the opinion of an individual known for his anti-environmental views; however, even among those who embrace the "personal virtue" of environmentalism, there is a tendency to think of the individual value—along with a relatively indistinct belief that there are positive environmental consequences associated with a given behavior—rather than the social value of such virtues. At least some contemporary environmentalism appears to be a lifestyle choice, with the emphasis on *style*. For example, the popular social networking site "Facebook" has an application that allows users to post a list of their "green leaves," essentially enumerating their individual attempts to be environmentally virtuous: "I do not shop at Wal-Mart"; "I do not commute"; "I recycle, even when it's not convenient"; etc. The authors of this application encourage potential users to "declare the ways in which you are green, or soon will be green!" In certain contexts, such declarations can indeed be useful, even virtuous. Sharing one's attempts to be virtuous with one's friends is certainly beneficial to both parties. Aristotle points out that friends help each other in their virtue. In addition, there is arguably some benefit in making a public show of one's virtue as a means of bearing witness to an injustice, as in the case of civil disobedience or conscientious objection to military service. Nevertheless, perhaps because of the detached context of social networking, it is hard to avoid the sense that part of the point is to wear one's environmentalism on one's sleeve, to bear witness to one's own virtue (in contrast, presumably, to folks who are not "green"), even to boast.[10] The questions that dominate environmental discourse in mainstream media also suggest that environmental virtue is a purely personal matter, and often a matter of "virtuous" consumption and style. Paper or plastic shopping bags? Locavore, vegetarian, or vegan? Organic cotton, hemp, or post consumer recycled designer clothing? WVO, biodiesel, ethanol, or hybrid gasoline? How can you "green" your overseas vacation? Which organic, boutique wine should you serve with the Chilean Sea Bass at your Earth Day party?

This is not to dismiss the importance of personal virtue—far from it—nor to imply that the examples given are not, in some measure, associated with environmental virtues. Personal virtue is both important for environmentalism and essential to flourishing. Rather, the point is to draw our attention to the fact that environmental virtue is frequently seen as (1) a personal preference rather than a requisite component of human flourishing and (2) a matter of individual actions isolated from social engagement. Both of these perceptions are errors. Something important is missing in this account of environmental virtue. In what follows, I want to look at public virtue and its role in environmental virtue. First, I'll try to outline, in broad strokes, different types of virtue in order to frame the discussion of public virtues and, in particular, a subclass of virtues I will refer to as political virtue. Second, I'll focus on practical problems and address the inadequacy of personal virtue for effecting social change and, therefore, for addressing most environmental

[10] Facebook, including the "green leaves," is in most cases not really a conversation between two genuine friends, in either the rarified Aristotelian sense of true friendship or even the contemporary, quotidian sense of friendship. Many facebook users have hundreds, even thousands, of "friends." Surely enumerating one's virtues in such a context has an element of braggadocio. Aristotle points out that real friendships are very rare (Aristotle 1925, pp. 197 and 243).

crises. Finally, I'll argue that public and political virtues are necessary, if under emphasized, conditions for the flourishing of the individual, and that they are important complements to more traditional environmental virtues.

Personal Virtue and Public Virtue

As Aristotle points out, it is difficult to fully capture some virtues and vices with a single term, which is why some virtues and vices lack colloquial names. In order to discuss such unnamed virtues, he tends to offer a sketch of the disposition in question by describing it and by contrasting it with opposing vices (Aristotle 1925, pp. 95–96 and 99). What Aristotle says of individual virtues is also, in this case, true of a class of virtues. Neither "public virtue" nor "political virtue" fully captures the scope of the characteristics with which we are concerned. Nevertheless, by offering a rough definition and then fleshing it out with some descriptive content, we can circumscribe each type in sufficient detail for the arguments that follow.

There is no sharp line distinguishing public virtue from personal virtue; because we are social beings, all virtues have both personal and public effects. The difference is a matter of emphasis rather than essence. Public virtues (1) contribute primarily to the well being of the community, though there may (2) be an element of direct personal benefit to having the virtue, and there is always at least some indirect personal benefit insofar as the individual is a member of the community. Examples of public virtues include neighborliness (proper goodwill towards members of one's community) and emplacement (attachment to and concern for one's local environment or place). Also among the public virtues are political virtues such as political engagement.[11] In contrast to public virtues, personal virtues contribute (3) primarily to individual well being, though there is usually (4) a secondary or indirect impact on community well being. Examples of personal virtues include temperance (a mean between excessive asceticism and gluttony) and economy (a mean between miserliness and profligacy). This is not to dispute that virtues contribute to personal flourishing, but to point out that different sorts of virtues have different objects. Virtues can be eudaimonistic without aiming at personal flourishing or, indeed, even aiming at the virtuous agent.

In light of these distinctions, another way to conceive of the difference between public virtue and private virtue is on the basis of the aim, or intention, of the virtue. When the primary aim or object of a virtue is other people or the community, we

[11] Political engagement serves as a virtue in at least two senses. First, with "engage" taken as a transitive verb, it the virtue of active concern for the political issues and well being of one's community. It indicates a mean between apathy (a lack of concern with those issues) and fanaticism (an intense and often irrational devotion to a particular perspective on those issues). Second, with "engage" taken as an intransitive verb, it is the virtue of active involvement with civic and political life. Here it indicates a mean between passivity (an unwillingness to act) and zealotry (a willingness to act precipitously or rashly). Although there is often an overlap of these virtues and vices, they are conceptually distinct. For example, it is possible to be a fanatic who is irrationally committed to an idea without being a zealot who acts in support of that idea. Likewise, it is possible to be engaged in the first sense but not in the second (e.g., an intellectually engaged professor who has thoughtful and educated positions on issues, but who does not act in support of her intellectual commitments).

can say that the virtue in question is a public virtue, even if there are personal benefits associated with it. When the primary aim of a virtue is personal, the virtue is a personal one, even if there are clear benefits to the community associated with individuals practicing the virtue in question. For example, in most cases, temperance is a personal virtue because its immediate goal is the flourishing of the individual. Temperate people consume the proper amount and so benefit themselves by avoiding health conditions associated with gluttony or excessive asceticism. However, temperate people also benefit their community (local and global) by consuming no more than what is appropriate, thereby helping to address communal problems like global warming, land use, and poverty. Another example is the virtue of neighborliness, which is a public virtue. People who are neighborly—a mean, perhaps, between being antisocial and being a busybody—have appropriately friendly and supportive interactions with others in their community and contribute to making that community more friendly and cohesive. Other people in the community and the community itself are the objects of the virtue of neighborliness. However, neighborly people also benefit themselves insofar as they develop relationships that support them in various ways and because it feels good to have connections to one's community.

The important thing to note is that environmental virtue ethics—in fact, virtue ethics in general—tends to focus on the direct or indirect benefit for the virtuous agent (2 and 3 above) to the neglect of the direct or indirect benefit to the community (1 and 4 above).[12] Is this legitimate for a virtue ethics that takes seriously the environmental aspects of human flourishing?

Public Virtue and the Practical Issue of Environmental Crisis

Environmental problems are often the result of an agglomeration of individually insignificant actions and, as such, they usually require solutions that address collective action. Collective action, in turn, requires political leverage. Thus, community relationships, institutional structures, and political will all have an important role to play in responding to environmental crises. Assuming that environmental crises are at least part of the matter at issue for environmental ethics, any environmental virtue ethic must help to address actual environmental crises and so must include the social and political aspects of environmental virtue. It would seem strange to articulate an environmental virtue ethics that, in its focus on personal flourishing, did nothing to address actual environmental problems.

[12] This is true whether the emphasis is on the benefit for the agent in question (e.g., cultivating a "virtue" solely for the benefits one derives from it) or on the benefit for individuals, generally speaking, rather than the community. In both cases, we tend to focus on the relationship between the individual and the environment, to the relative neglect of the relationship between the individual and the community and the relationship between the community and the environment. This is not usually an issue of *omitting* attention to (or cultivation of) a certain class of virtues in either the popular or the academic sphere. Rather, the issue has to do with what sorts of virtues are *emphasized*. Eudaimonism, to which I subscribe, tends to focus on individual flourishing and the relationship of the individual to the environment. However, our current social, economic, and environmental contexts should, if anything, cause us to *emphasize* the public and political virtues that are at issue here.

The necessity of relating politics to virtue should not be surprising; the connection between the two has long been recognized. *The Nicomachean Ethics*—the very text that founded virtue ethics, at least in the West—is a preamble of sorts to a treatise on politics (Aristotle 1925, pp. 269–276). Plato famously draws an analogy between the city and the soul in *The Republic*, asserting that a well functioning soul and a well functioning city closely resemble each other (Plato 1968, p. 113). Building on this analogy, we might say that a purely personal virtue ethics is as incomplete as an isolationist political program. Just as it would be foolish to run a state as if its only concerns lay within its own borders, it is unwise to think of virtue as something concerned only with the personal well-being of the individual. Nations inevitably have relationships with—and, therefore, rights, duties, and obligations to—other nations. Likewise, individuals always exist in a community—indeed, simultaneously within a number of overlapping communities—which shape their rights, their obligations, and their virtues.[13] International relations influence the domestic well-being of a state, and social and political virtues shape the personal well-being of the individual. "No man is an island, entire of itself" (Donne 2008, p. 62).

However, despite philosophical precedent and the recent interest of some virtue ethicists, the connection between the individual and the community remains underappreciated. Certainly the importance of political virtue for individual flourishing is insufficiently emphasized. This lacuna extends to, and is particularly glaring in, the area of environmental virtue ethics, precisely because environmental crises depend on collective rather than individual action. It is for this reason that Robert Hull argues that the looming question for environmental virtue ethics is how it might be coordinated with political philosophy. He gestures toward the importance of political virtues for answering this question when he suggests that "*activism* [is] a central avocation of good ecocitizenship" (Hull 2005, p. 108).

Personal Virtue in the Commons

The idea that virtue ethics can help to solve environmental crises through the power of virtuous examples is an attractive one.[14] When one person, or a small group of people, improves her life (i.e., flourishes) by riding rather than driving, shopping locally, gardening, or restoring a local ecosystem, others are influenced by her example. We prod or inspire others to environmentally virtuous behaviors not only through confrontation, argumentation, and litigation, but also through the power of examples. Nevertheless, there are many problems that, if not *literally* insoluble by inducing virtuous action with virtuous examples, are *practically* so. Environmental

[13] I am not making the obvious point that, from the perspective of virtue ethics, flourishing and the virtues that constitute it are, to some extent, socially constructed. Rather, the point is that there are virtues that are directed at the well functioning of the community or social group rather than the flourishing of the individual (although the individual may well flourish as a result of practicing those virtues).

[14] Indeed, I myself am interested in the uses of virtue ethics for environmentalism, and I support this belief in the power of virtuous examples. See Treanor 2008.

crises may be local (e.g., urban sprawl, lack of green space, etc.), regional (e.g., water use in the American West), or global (e.g., anthropogenic climate change, the changing pH of the ocean, etc.); however, relatively few environmental crises are individual problems that can be adequately addressed by one person, even one virtuous person. I can stop littering, but I cannot, alone, end littering in my neighborhood. No matter how committed I am to a bicycle lifestyle, air quality will not change in the Los Angeles basin because my family and I choose to bike rather than drive around the city. My attempts to live locally will not change American dependence on a far-flung distribution system based on fossil fuels. It is hard to see how individual virtuous actions, on their own, will change broad-based social problems.

The problem is that actions that are individually insignificant and therefore not particularly vicious are, when multiplied across a population and combined in effect, both vicious and environmentally devastating. Garret Hardin addressed this issue in his justifiably famous essay, "The Tragedy of the Commons" (Hardin 1968). Hardin's initial claim is that certain problems—among them the "population problem," and, I would argue, most other environmental crises—have "no technical solution" (Hardin 1968, p. 1243). Although technological innovation will certainly help us in a variety of ways, we cannot finally and decisively invent our way out of certain environmental crises. What, then, are we to do? One possibility is that responding to such challenges will require a new way of being rather than new, "greener" technology. Such a view appears congenial to virtue ethics insofar as it suggests that we need a new vision of flourishing and a new *modus vivendi* rather than a technical fix. Sandler, for example, suggests "one plausible way to raise the salience of environmental issues is to emphasize how they are implicated in the quality of our own lives, as well as in the many things we care about" (Sandler 2007, pp. 116–117). I agree that virtue oriented arguments and the examples of virtuous persons can offer powerful models that can influence, sometimes dramatically, a community's view of "the good life"; indeed, I think we could be *more* emphatic about the potential efficacy of virtue in this regard.

However, while virtue theory no doubt has something to contribute to the analysis of ethical behavior in a commons, Hardin's account should give us pause and check unjustified exuberance regarding the efficacy of virtue discourse, especially when articulated in a eudaimonistic model that emphasizes personal flourishing. It is true that virtue oriented arguments and environmental exemplars will sway *some* people, but are they likely to sway *enough* people? Aristotle cautions us about the efficacy of arguments. "If arguments were in themselves enough to make men good, they would justly… have won very great rewards…. [However,] while they seem to have power to encourage and stimulate [a *few*], they are not able to encourage the *many* to nobility and goodness" (Aristotle 1925, p. 270). The main point of "The Tragedy of the Commons" is that rational actors will defect—that is, act in a vicious manner—in an unrestricted commons. Defection from the common good in favor of personal good is the "rational" thing to do. It is true that some people might—for any number of reasons, including virtue—resist the temptation and act against their own personal interests out of concern for the common interest. Hardin, however, dismisses the long-term impact of such virtuous people and, by extension, the virtuous examples they set. Self-sacrifice in a commons is self-defeating.[15]

Indeed, in terms of the population problem (Hardin's concern), virtue is literally self-eliminating. Because children tend to live lives that resemble their parents', people who choose to have fewer children, or none at all, will sooner or later be out-competed and outnumbered by those who choose to have large families. Although Hardin's account is overly simplistic—e.g., there are many other factors influencing family size—he does have a point about the relative competitive fitness of ideas. It is not enough to set a good example or make a good argument for, say, the virtue of moderation if that argument cannot compete in the marketplace of ideas. A quick internet search of popular television shows reveals: *Privileged*, *Gossip Girl*, and *90210* (all about wealthy teenagers); *Who Wants to be a Millionaire*?, *Deal or No Deal*, *Are You Smarter than a 5th Grader*?, and *The Moment of Truth* (vacuous shows, some designed to humiliate the contestant, about winning money); and *Survivor*, *Big Brother*, and *The Apprentice* (so-called "reality" shows encouraging positively vicious behavior in the pursuit of cash prizes and instant celebrity). Television, a major influence on contemporary social values and mores, touts shows about excess and vicious behavior, not moderation and virtuous behavior. Pitching a new reality show about excess would, I imagine, have a relatively high level of success compared to a similar pitch for a show about moderation.

Returning to the issue of individual virtue in the commons, one has to wonder about the competitive fitness of the example made by one virtuous person, even a quite famous person, against the examples prevalent in the multifaceted, multinational, 24/7 on demand, juggernaut of advertising and entertainment. Seen in this context, an exhortation to virtue seems suspiciously similar to an appeal to conscience. Both implore us to "do the right thing"—e.g., be honest, moderate, or humble—in a cultural context that suggests manipulating the truth, immoderation, and self-aggrandizement lead to happiness. Those susceptible to such appeals will alter their behavior in ways that appear to make their lives more, not less, difficult and less, not more, socially accepted. In contrast, those impervious to such appeals can forge on engaging in behavior that is, if not socially lauded, at least socially sanctioned and can reap any benefits such behaviors offer. Because of the problems with appeals to guilt, conscience—or, I would add, virtue—Hardin suggests two more efficacious ways of addressing threats to a commons: (1) privatization and (2) "mutual coercion mutually agreed upon" (Hardin 1968, p. 1247) in other words, laws.

Does this mean that virtue has nothing to contribute to solving the tragedy of the commons? No. First, as any virtue ethicist would know, and as several economists have pointed out, sustainability is possible in the context of a sufficiently small commons used by a sufficiently virtuous population (Ostrom 2001). However, in addition, virtue theory must engage Hardin's proposed solution to the tragedy because laws do not write, pass, or enforce themselves. Bracketing for the moment the possibility of a grassroots revolution in virtue, any solution to the tragedy of the

[15] "Conscience [particularly in the case of the population problem] is self-eliminating" (Hardin 1968, p. 1246).

commons is going to require the political will to legislate the protection of the commons. Passing laws will require the political virtue of engaged activists and politicians. If we are to avoid trading one problem (i.e., environmental degradation) for another (e.g., corruption or political oppression), laws protecting the commons must be crafted and enforced in the right way. Here again, we need men and women with highly developed political virtues. Environmentalism needs politically active advocates at every level: local, regional, national, and international. It needs advocates who formally engage the political process as candidates, campaign workers, and environmental lobbyists. It also needs advocates whose political contribution is less formal, such as authors, political commentators, and educators. Without these people, environmentalism—especially in a democracy—is doomed.

Political Virtue and Environmentalism

Public virtues are character traits that bring us into virtuous relationships with our communities and environments. Although there are many public virtues that ought to play a prominent role in environmental virtue ethics, the subset of political virtues is especially worthy of our attention. Political virtue is actually a type or class of virtues rather than a singular virtue. Indeed, in the broadest sense, all public virtues are political to one degree or another. Nevertheless, it is useful to maintain a distinction between political virtues broadly speaking (i.e., most, if not all, public virtues) and political virtues more narrowly construed (e.g., political engagement), and it is the latter that we will refer to as "political virtue" here. The politically virtuous person is, first of all, a person who is engaged intellectually (e.g., thinking reasonably about the issues, educating herself about the issues, discussing the issues with others) and practically (i.e., someone who will act in support of her reasoned convictions). Such a person cares about her community and seeks to better understand how it can flourish. Moreover, her concern is not passive; she is actively engaged in bringing about the conditions for that flourishing. In an environmental virtue ethics, both "well-being" and "community" should be taken expansively, the former to include all the things (material, social, aesthetic, etc.) that allow members of the community to flourish and the latter to include all the interrelated members of the community: other persons, groups of persons, animals, plants, and biotic communities.

One might argue that political virtues are well and good, but that they are distributed virtues that some members of the community must possess and cultivate rather than universal virtues that every member of the community must possess and cultivate. Clearly there are some virtues that are role-specific; however, political life is not something that should be limited to an elite few, but something in which all persons should be engaged. True, not everyone will participate in the same way or to the same degree. Some people will devote their lives to politics, serving as politicians, advisors, or advocates; for such people, political virtues will play a particularly prominent role in *eudaimonia*, as courage is a particularly important virtue for a soldier. Nevertheless, just as all persons, not just soldiers, should cultivate courage, all persons should cultivate political virtues. This is especially true if we take a more expansive view of political and civic life, one that can accommodate degrees of political engagement at a variety levels (i.e., local, regional, or national). Conceived in this way, it seems clear that everyone ought to participate in political and civic life in some measure. Indeed, Tocqueville thought that one of the strengths of American democracy (and of American culture) that it affords, or afforded, so many opportunities for participation in civic and political life (Tocqueville 2001). A life without some minimal exercising of political virtues is impoverished.

Given this framework, political virtue can manifest itself in diverse ways that correspond to our diverse constitutions. Therefore, while we should all "find sustainable ways to engage in politics," we should, ideally, "find political roles that we enjoy" (Cafaro 2005, p. 153). A community organizer, a local politician or city council member, a political volunteer, a person who organizes or participates in demonstrations and rallies, and an active member of local organizations dedicated to improving the community would each be engaged in activities conducive to cultivating political virtues. In a broader sense, organizing neighborhood recycling, composting, or freecycling, starting community gardens, authoring amicus briefs or careful comments on local EIRs, writing newspaper editorials, authoring blogs, and lecturing to the eco-curious are all opportunities to cultivate politically virtuous characters. Virtues or character traits exhibited in some of these activities include patience, sensitivity, self-sacrifice, care, commitment, neighborliness, community concern, humility, learning, and wisdom, among others. The focus of one's involvement could be global, national, or local, though some of the beneficial aspects of political virtue discussed below are more easily seen in the case of local participation.

The idea of political virtue is fairly intuitive; however, in terms of actual experience, we are more familiar with political vice. There are many ways to fail to be virtuous in the pursuit of political ends, among these are fanaticism and apathy. The former can manifest itself in various ways, but "ideology" and "utopia" offer two useful illustrations of fanatical pathologies in the context of environmentalism. Ideology and utopia are the two poles of what Paul Ricoeur calls the "social imaginary," the set of imaginative practices by which we define our social, political, and cultural communities (Ricoeur 1991, p. 181). While both are necessary, an imbalance in favor of either one produces a vicious disposition. Ideology is "tied to the necessity of any group to give itself an image of itself, to 'play itself' in the theatrical sense of the word, to put itself at issue and on stage" (Ricoeur 1991, p. 182). There is no social group that does not relate to its own being, as least in part, through such a representation of itself. Social groups are an effect of this symbolism. Ideology integrates and unifies a social group via interpretation of relevant symbols. However, the omnipresence and necessity of ideology does not remove the possibility of vicious manifestations of ideology; history is littered with examples of such distortions. The pathological potential in ideology stems "from its function of reinforcing and repeating the social tie in situations after the fact," which can result in "simplification, schematization, stereotyping and ritualization [growing] in the distance between real practice (event) and interpretation" (Ricoeur 1991, p. 182). This pathology is especially acute when ideology is co-opted by systems of authority. "The function of dissimulation clearly wins out over the function of integration when ideological representations are captured by the system of authority in any given society. All authority seeks to legitimate itself" (Ricoeur 1991, pp. 182–183).

Unlike ideology, utopia does not engage in dissimulation; it knows and claims itself to be utopian, the imaginary project of another reality (Ricoeur 1991, p. 183). Because there are many different sorts of utopias, we cannot define utopia in terms of its content. Rather, we need to focus on the specific function of utopia, resisting, as with ideology, the temptation to simply reduce it to its pathological manifestations (Ricoeur 1991, p. 184). The central idea of utopia is the no-place (*u-topos*) implied by the word itself, which allows the field of the possible to extend beyond the real. Utopia allows us to imagine other realities, other ways of living. It allows us to rethink "the nature of family, consumption, government, religion, and so on" (Ricoeur 1991, p. 184). However, in imagining these alternative societies, the pathological aspect of utopia becomes evident. Arising from a nowhere, utopias have a tendency to

> subordinate reality to dreams, [fixate on] perfectionist designs, and so on. [There is, on some accounts, a sort of] schizophrenia at work here, a logic of all or nothing, ignoring the work of time… disdain for the intermediate stages and an utter lack of interest in taking a first step in the direction of the ideal… blindness with respect to the contradictions inherent in action… nostalgia for a lost paradise concealed under the guise of futurism… (Ricoeur 1991, p. 185).

Utopia, then, is the counterpoint to ideology. It is the *yang* to ideology's *yin*. Where the latter is a force of social integration (and cohesion), the former is a force of social subversion (and critique). Like *yin* and *yang*, ideology and utopia are in constant, interdependent movement against each other. Thus, for example, utopias not only critique and subvert power, they offer alternative ways of organizing and exercising power. A politically virtuous person will be on her guard against both fanatical commitment to the status quo and fanatical insistence on impossible dreams. Although ideology and utopia differ and are, to an extent, defined over and against each other, the vicious manifestations of either one stem from something like fanaticism or zealotry.[16]

If fanaticism is one way to fail to be politically virtuous, the contrary failure can be seen in apathy or passivity. A politically apathetic person is the sort of person who simply does nothing to care for her community, either out of a lack of concern (apathy) or due to an unwillingness to become involved and act (passivity). In the former case, the apathetic person may have no connection to the community (e.g., an itinerant person) or she may have a more generalized apathetic outlook. Whatever the cause, lack of concern for the community leaves her content to protect her personal happiness isolated from her neighbors (to the extent that this is possible). In contrast, passivity is apparent in people who are concerned about their communities, but who do not act on that concern in any meaningful way; they worry

[16] Of course, in some contexts, it may be the case that political virtue is closer to the vice of zealotry than to the vice of apathy. As renowned environmentalist and political activist David Brower once quipped, "I was not always unreasonable, and I am sorry for that." On the distinction between fanaticism and zealotry, see note 11 above.

and complain but do not act. Apathy and passivity are political vices because they do nothing to actively promote the good. Political virtue requires *action*, which is why Thoreau has such a dim view of mere voting.

> All voting is a sort of gaming, like checquers or backgammon, with a slight moral tinge to it, a playing with right and wrong, with moral questions... The character of the voters is not staked. I cast my vote, perchance, as I think right; but I am not vitally concerned that that right should prevail. I am willing to leave it to the majority.... Even voting for the right is doing nothing for it (Thoreau 1973, pp. 69–70).[17]

Addressing environmental crises will require people with public virtues, especially political virtues, and the mettle to act on these virtues. Atomistic individuals cannot solve environmental crises; the commons is ruined even if a significant percentage of its inhabitants live sustainably. A small percentage of defectors from the environmental-social contract put into motion the inexorable mechanisms that lead to tragedy. Unless we can identify a constitutionally grounded principle to protect the environment, or amend the constitution to establish such a principle, the only way to establish the political and legal "coercion" necessary to protect the environmental commons is through "mutual agreement."[18] Mutual agreement, however, will require adroit political maneuvering at every level. Environmentalism must come to be thought of in terms that are explicitly and unavoidably political. We can no longer afford to think of environmentalism solely, or even primarily, in terms of personal virtues. It is attractive, no doubt, to think that one can, through the cultivation of virtue and by dint of hard work, escape the dilemma in which we find ourselves, that one could escape the rapidly swamping ship and set sail on one's own lifeboat. But that is ultimately a dangerous fantasy. We're all in this together. In a commons, either we are all saved or none of us is saved.

We need to free ourselves from our frontier past and its virtues, not so that we can abandon them entirely, but so that they no longer exert a tyrannical hold on us. One of the most powerful insights of virtue ethics is the contextual nature of ethical action. Environmentalists need to acknowledge the momentum—political, economic, social, cultural, and technological—is toward the increasingly rapid destruction of the global commons. Individual virtuous behaviors—while admirable and, to a degree, effective—are inadequate to fight the battle we have before us. In the contemporary environmental context, political engagement is at least as important as simplicity, arguably more so. Today environmentalists need a bit less communion with nature and a bit more political organizing, a bit less Henry David Thoreau and a bit more David Brower.[19]

[17] Thoreau goes on to exhort us to "cast your whole vote, not a strip of paper merely, but your whole influence" (Ibid., 76).

[18] There are thinkers who have suggested the possibility of pursuing environmental rights on the model of civil rights (which, presumably, therefore would be codified like civil rights). See, for example, Bullard 2008.

[19] I must admit that it pains me to write these words. I confess that I remain enamored by "frontier" virtues. Perhaps more than most of my environmental friends and colleagues, am seduced by the idea of an "escape" to some form of independent self-sufficiency in the wild. My own life and my own

Public Virtue and *Eudaimonia*

However, the good news is that political virtue does not frustrate a eudaimonistic approach to virtue ethics. Not only are public and political virtues a necessary component of any effective response to environmental crises, they are a necessary component of individual flourishing. As Aristotle points out, political association is an essential human activity—"man is by nature a political animal" (Aristotle 1946, p. 5). Outside the *polis*—that is, outside our social and political relationships with other people—we are no longer truly human. "The man who is isolated—who is unable to share in the benefits of political association, or has no need to share because he is already self-sufficient—is no part of the *polis*, and must therefore be either a beast or a god" (Aristotle 1946, p. 6). One cannot flourish as a human being without flourishing as a member of a social and political community. Note that this is not the commonplace environmental observation that one cannot flourish in an environment or community that is so degraded as to preclude flourishing—"there is no flourishing to be done on a dead planet."[20] It is true that we cannot fully flourish in a severely degraded environment; but the point here is different. We cannot fully flourish unless we flourish *as members of a community*, whether or not the community and environment in which we live is healthy. Flourishing cannot take place in isolation because we need social relationships, we need to contribute to a community, and we need to care for a place in order to fully flourish.

While it is true that the community is the intentional object of political virtues, such virtues also contribute to personal flourishing. People who exhibit virtues such as political engagement are more likely to feel that they have some control or influence over their community and environment. Apathy, which we said is one way to fail with respect to political virtue, is deeply corrosive to individual flourishing. "Giving into apathy means acquiescing in powerlessness. It means allowing others to circumscribe your life and your children's lives"; social or political engagement on behalf of the environment and the life you want "engages a basic human capability for political action" (Cafaro 2005, p. 152). People engaged in shaping their community and environment feel that they have some influence over their quality of life, which is empowering and satisfying. In addition, politically virtuous people have a sense of contributing to something larger than themselves. Another

Footnote 19 continued
environmentalism has, to a large degree, been shaped by this romantic notion. Though I admire Brower—who was himself inspired to environmentalism by a youth spent backpacking and climbing in the Sierra Nevada—my own heart and imagination are with Thoreau and Muir. Moreover, I do not repudiate this idea or the virtues that go with it, at least not completely. My argument here is that we must not remain *captive* of these notions of virtue and flourishing in a context that has radically changed. Bill McKibben illustrates, in the Introduction to the new edition of his groundbreaking *The End of Nature*, the necessity of getting beyond the idea of individual answers to the environmental crisis. "[Writing *The End of Nature*] in 1989, with the solipsism of someone in his mid-twenties, I focused on individual human efforts—smaller families, reduced consumption, and so on. I still think these are important, but I've come to think that equally important changes lie elsewhere: in the direction of stronger, tighter communities" (McKibben 2006, p. xxiii). McKibben's recent work, political and academic, bears the mark of this new orientation (see 350.org and McKibben 2007).

[20] A play on David Brower's clever remark, regarding business ethics, that "there is no business to be done on a dead planet."

way of stating this is that engagement contributes to the meaningfulness of one's life. Without meaning, life is not worth living. Of course, political-environmental engagement is not the only source of life's meaning, even for most environmentalists. Nevertheless, this is an important consequence of social and political engagement. Thus, while political engagement and the virtues constitutive of it have as their intentional object the community or environment, they contribute in very concrete ways to personal flourishing.

Other Virtues Associated with Political Virtue

In addition to the direct benefits of virtuous political engagement, exercising political virtues cultivates and supports other virtues, which, in turn, support political virtue. However, while this synergy is noted by some environmental virtue ethicists, the importance of these connections is underemphasized. It is in this vein that Hull laments that environmental virtue ethicists "tend to reduce human excellence to green virtue or to ignore or misinterpret its connection to other excellences" (Hull 2005, p. 100).[21] Schellenberger and Nordhaus make a similar criticism of environmentalism as a whole. Environmentalists have an overly narrow vision of what constitutes an "environmental" problem, which causes them to miss the forest for the trees.[22] We can talk about virtues in isolation and it is certainly possible to have one virtue (e.g., simplicity or economy) without having other virtues (e.g., courage); nevertheless, in practice many virtues support each other in ways that are instructive. On reflection, this synergistic relationship makes sense, given that each individual virtue contributes to flourishing. Whether or not we embrace Aristotle's suggestion that there is a unity of the virtues, is seems reasonable to address the ways that certain virtues support and nourish other virtues (Aristotle 1925, pp. 156–158). A partial list of virtues or dispositions synergistically related to political engagement would include hope, friendliness, truthfulness, open-mindedness, justice, benevolence, connection to community, connection to place, diligence, and commitment to a cause larger than oneself. A complete analysis of the relationship between political virtue and other virtues would require a much longer treatment; therefore, for the purposes of illustration, let's look at the virtues of benevolence and hope.

Benevolence

Benevolence is, perhaps surprisingly, associated with social and political engagement. This may seem counterintuitive given our experience with "culture wars" in politics over the past decade, and indeed may seem an unlikely virtue to be

[21] Aristotle goes so far as to suggest a "unity of the virtues" under the rubric of phronesis in *The Nicomachean Ethics* (Book VI, 13). Although Sandler points out the importance of "virtues of environmental activism," including "character traits such as commitment, astuteness, discipline, attentiveness, discernment, fortitude, creativity, courage, self-control, cooperativeness, patience, solidarity, perseverance, and optimism" (Sandler 2007, p. 49).

[22] See, for example, Schellenberger and Nordhaus 2008.

associated with politics in any context.[23] After all, we are apt to think of politics in Machiavellian terms, accepting Clausewitz's assertion that war and politics are two manifestations of the same impulse. However, the assertion here is that benevolence is a virtue synergistically related to political *virtue*, not necessarily to politics as commonly practiced. If "benevolence" sounds entirely too utopian a term to apply to politics, perhaps "friendliness" carries marginally less objectionable connotations. Aristotle, for example, suggests that friendship and politics are related, calling friendliness one of the virtues of social intercourse. It is the mean between obsequiousness (i.e., praising everything) and churlishness or contentiousness (i.e., criticizing everything); it helps us to "associate with people in the right way" (Aristotle 1925, p. 99) and so facilitates the functioning of a social group by maintaining proper relationships among members of that group. Friendship—and, to a lesser degree, friendliness—fosters a kind of fellowship that is the basis for community and social interaction. Therefore "friendship seems to hold states together" (Aristotle 1925, p. 192).[24] Friendliness is also part of Geoffrey Franz's account of benevolence.

> *Benevolence* as such is a genus or family of virtues that involve a direct concern for the happiness and well-being of others. Virtues of benevolence include compassion, friendliness, kindness, and generosity. Feelings of affection need not be present for there to be benevolence, though the two traits are often found together (Franz 2005, p. 123).

Benevolence clearly supports other public virtues like neighborliness (indeed, neighborliness is in large part benevolence applied to one's community) and emplacement. A benevolent person is more likely to be tied to her community and her place, and a person tied to her community or place is more likely to be benevolent. Benevolence also supports specifically political virtues, because a person who feels attached to her neighbors and to her place is more likely to be interested in the well being of her community. And, crucially, a person with the virtue of benevolence is more likely to *act* on her concern for her community (Franz 2005, p. 124).[25] It was suggested above that many of the beneficial aspects of political engagement are more evident when that engagement takes place at the

[23] Not all political engagement encourages or exhibits benevolence. It seems clear that our current political climate—dominated as it is by spin culminating in outright lies, AM talk radio that doesn't even rise to the level of sophistry, "swift boat" attacks that target irrelevant aspects of candidates' records (including the vices of their extended families), and similar tactics—is more characterized by churlishness or contentiousness. However, although people from both ends of the political spectrum contribute to this phenomenon, complaints about the state of politics indicate that many people realize that this type of politicking is vicious.

[24] While this is no longer strictly true due to the size of the *polis*—the United States having recently passed the 300,000,000 mark—the underlying insight, that friendliness and friendship are the basis for *communitas*, remains true. Think of the difference between communities where neighbors are friendly with each other and "communities" where neighbors are indifferent or hostile to each other. The former groups exhibit a cohesiveness that makes them real communities; the latter concatenations of people hardly deserve the name community.

[25] Aristotle agrees, at least concerning friendliness. "It implies no passion or affection for one's associates" (Aristotle 1925, p. 99).

local level rather than the national or international level. This is clearly the case with benevolence, for it is easier to cultivate and exercise compassion, friendliness, kindness, and generosity in a local context.

Although benevolence is a public virtue—its intentional object is other individuals, the community, or the environment—there are substantial personal benefits associated with benevolence. Friendliness and friendship are both among the virtues of benevolence, and Aristotle clearly and convincingly ties both these virtues to personal flourishing. First, friendliness improves one's chances of developing "friendships" of utility and pleasure because the friendly person is very like a good friend, minus the actual affection (Aristotle 1925, p. 99). So, friendly people are likely to have an easier time in life, having as many "friends of utility" as necessary, and more likely to have a pleasant life, having sufficient "friends of pleasure." While none of these associations are as important as "true" friendship for Aristotle, they are conducive to flourishing. Moreover, it seems reasonable to assume that friendliness, while not strictly necessary, is influential in actual (true) friendships as well. Aristotle reserves a full fifth of the *Nicomachean Ethics* for the discussion of friendship, which is "necessary" for flourishing and without which "no one would choose to live, though he had all other goods" (Aristotle 1925, p. 192).[26] Friendliness is only one aspect of benevolence according to Franz, but within this one element we have found strong reasons to conclude that benevolence contributes in very concrete ways to personal flourishing.

Hope

A second virtue that is synergistically related to public and political virtue is hope, a virtue that is absolutely essential to *eudaimonia*, especially as articulated in environmental virtue ethics. People who are politically engaged and involved in their communities are more likely to be hopeful that things can improve. Involved people are more likely to feel that they can influence their community and make a positive impact on the environment in which they live. If one feels one can make a positive impact, *ipso facto*, one must believe that things can improve. The synergistic relationship between political virtues and hope flows the other way as well. Hopeful people, people who believe in the possibility of things improving, are more likely to cultivate and develop public and political virtues because they have reason to believe that exercising such virtues can make a difference. If one does not believe that things can improve, one is very unlikely to attempt to improve them. Thus, hopeful people are more likely to be engaged members of their communities, cultivating various public and political virtues; and being an engaged member of the community, cultivating public and political virtues, is likely to make one more hopeful.

However, it might be objected that political engagement and deeper awareness of environmental crises is more likely to lead to despair than hope. The contemporary

[26] He continues, "…in poverty and in other misfortunes men think friends are the only refuge. It helps the young, too, to keep from error; it aids older people by ministering to their needs and supplementing the activities that are failing from weakness; those in the prime of life it stimulates to noble actions… for with friends men are more able both to think and act" (Ibid.).

environmental situation does not seem to counsel hope; it would be easy to become overwhelmed by the increasingly dire and urgent threats to the environment and the substantial obstacles to any effective response. Most of the major emitters of CO_2 refused to sign on to the Kyoto Protocol, and those that did are failing to meet their targeted reductions in emissions. It seems unlikely that this year's climate meeting in Copenhagen will be able to produce an agreement on drastic, enforceable cuts sufficient to avoid a climate catastrophe. Fisheries around the world are in decline from overuse. Population continues to expand. American-style consumption is being successfully exported to cultures around the world. And, finally, in the face of these threats, and others, there is no effective international body for establishing and, crucially, enforcing environmentally ethical behavior. The situation does not seem to offer much ground for hope. As Gabriel Marcel notes, in a different context to be sure, "despair is possible in any form, at any moment and to any degree, and this betrayal may seem counseled, if not forced upon us, by the very structure of the world we live in" (Marcel 1995, p. 26). In the face of this complex and daunting constellation of crises it would be easy to despair, to throw in the towel, acknowledge the cause is hopeless, and get what enjoyment we can out of life for as long as we can get it.

The contemporary environmental situation tempts us to, and may even "counsel," despair; but despair is fatal to both environmental progress and individual flourishing, and is therefore a vice. It is fatal to environmental progress because while it is true that we may not be able to adequately respond to certain crises in time to avoid their negative effects, failing to try ensures failure and often exacerbates the situation. The surest way to guarantee that burning of fossil fuels continues to wreak havoc with our climate is to believe this is inevitable and to accept it as *fait accompli*. Despair is fatal to flourishing because it undermines our belief in the significance of our actions and our lives. If we come to believe our lives are insignificant and do not matter, we cannot flourish in any significant sense. If our lives do not *matter*, if they are, in the grand scheme of things, meaningless, then there is no real reason to do anything. Although despair is a possible reaction to the current environmental and political situation, it is not a reaction conducive to flourishing. As William James points out in *The Will to Believe*, temperamental optimism, which is certainly a kind of hope, is our best defense against our fears that "the ultimate mystery of things works [out] sadly," a possibility to which "the whole army of suicides" attests (James 1956, pp. 32 and 27).[27]

While both personal and public virtues ultimately contribute to one's flourishing, virtue ethics, including environmental virtue ethics, has tended to focus on the former to the neglect of the latter. Although this is a matter of emphasis rather than omission, it remains a grave problem. This is true not only because any account of flourishing that does not fully take into account the public components of human well being is bound to be wanting, but also because environmental virtue ethics is concerned with more than personal flourishing. Environmental ethics, whether or not it is based in virtue theory, must address the dangerous and widespread

[27] Marcel (1995) believes that the possibility of despair is the precondition for hope; if we could not despair there would be no need for hope.

challenges to environmental integrity that threaten our well being and the well being of the planet, and that frame the ethical situation in which we find ourselves.

Virtue ethics has long acknowledged the importance of context. Today, the need for political activism is so urgent, so dire, that it fundamentally changes the kind of person we need to be in order to flourish. Public virtues have always been and will always be an essential component of *eudiamonia*; however, the importance of these virtues—in particular political virtues—is heightened given the challenges of our time. First, climate change is urgent enough that most of the people reading this article will live to see at least some of its devastating effects. As I write this, numerous studies are concluding that the 2007 report of the IPCC was too conservative, and that climate change is proceeding more dramatically and more quickly than previously suspected. Its impacts will directly affect the ability of people to flourish. Moreover, if unchecked, the impacts of climate change will be even more serious for subsequent generations, including children born in the early part of this millennium. This is significant because our flourishing is in fact influenced by the fate of those living after we have died.[28] One need not be religious to feel deeply wounded by the possibility that the planet, its civilizations and cultures, and its people will suffer widespread and catastrophic damage in the years immediately following one's death.

Any response to the challenges that confront us will have to address the social and political structures within which that response will be developed. This is true at the national level because many of the people most responsible for environmental degradation live in democracies. It is also true at the international level because solutions between nations will require adroit political maneuvering to achieve environmental goals. Public and political virtues are essential to environmental virtue ethics because addressing environmental crises will require people who have developed these virtues to a high degree. Moreover, these virtues are necessary for all people at some minimum level because, while public virtues such as political engagement and neighborliness are not sufficient conditions for personal flourishing, they are necessary conditions for it.

I've heard many environmentalists hotly debate the question of whether or not one must be a vegetarian in order to be an environmentalist. This is indeed an important issue; however, political activism is a much more important and urgent issue. This is so much the case that I would go so far as to claim that, in the current environmental context, one cannot legitimately call oneself an environmentalist without conscientiously cultivating and exercising the public and political virtues that are *essential* to meeting the challenges that confront us.

[28] Aristotle claims that the good or bad fortune of our friends and descendents "seem to have some effect on the dead," but concludes that the effects are "of such a kind and degree as to neither make the happy unhappy nor to produce any other change of the kind" (Aristotle 1925, p. 23). However, it seems more probable that the likely fate of those who will outlive us or come after us will impact our happiness and flourishing while we still live. That is to say, the possibility or likelihood of one's friends and children suffering as a consequence of one's actions (or inaction) is likely to have some impact on one's well-being. It would be difficult to really flourish if you suspected that you were part of the last generation to participate in a certain culture. Imagine, for example, being the last speaker of a dying language.

References

Aristotle (1925). *Nicomachean ethics*. (trans: David Ross). Oxford: Oxford University Press, 1925.
Aristotle (1946). *Politics*. (trans: Ernst Barker). Oxford: Oxford University Press.
Bullard, R. D. (2008). Overcoming racism in environmental decision making. In L. P. Pojman & P. Pojman (Eds.), *Environmental ethics: Readings in theory and application*. Belmont, CA: Thompson Wadsworth.
Cafaro, P. (2004). *Thoreau's living ethics: Walden and the pursuit of virtue*. Athens and London: The University of Georgia Press.
Cafaro, P. (2005). Gluttony, Arrogance, Greed, and Apathy: An Exploration of Environmental Vice. In R. Sandler & P. Cafaro (Eds.), *Environmental virtue ethics*. New York: Rowman and Littlefield.
Dagger, R. (1997). *Civic virtues: Rights, citizenship, and republican liberalism*. Oxford: Oxford University Press.
Donne, J. (2008). "No man is an Island". In *Devotions upon emergent occasions*. Whitefish, MT: Kessinger Publications.
Franz, G. (2005). Benevolence as an environmental virtue. In R. Sandler & P. Cafaro (Eds.), *Environmental virtue ethics*. New York: Rowman and Littlefield.
Hardin, G. (1968). The tragedy of the commons. *Science*, *162*(3859), 1243–1248.
Hardin, G. (2008). Who cares for posterity? In L. P. Pojman & P. Pojman (Eds.), *Environmental ethics: Readings in theory and application*. Belmont, CA: Thompson Wadsworth.
Hull, Robert. (2005). All about EVE: A report on environmental virtue ethics today. *Ethics and the environment, 10*(1), 89–110.
James, W. (1956). *The will to believe*. New York: Dover.
Marcel, G. (1995). On the ontological mystery. In *The philosophy of existentialism*. (trans: Manya Harari). New York: Citadel Press.
McKibben, B. (2006). *The end of nature*. New York: Random House.
McKibben, B. (2007). *Deep economy: The wealth of communities and the durable future*. New York: Holt.
Ostrom, E. (2001). Reformulating the commons. In J. Burger, E. Ostrom, R. B. Norgaard, D. Policansky, & B. D. Goldstein (Eds.), *Protecting the commons: A framework for resource management in the Americas*. Washington: Island Press.
Plato (1968). The republic. (trans: Allan Bloom). New York: Basic Books.
Ricoeur, P. (1991). From text to action: Essays in hermeneutics II. (trans: Kathleen Blamey & John B. Thompson). Evanston, IL: Northwestern University Press.
Sandler, R. (2007). *Character and environment: A virtue-oriented approach to environmental ethics*. New York: Columbia University Press.
Schellenberger, M., & Nordhaus, T. (2008). *Breakthrough: From the death of environmentalism to the politics of possibility*. New York: Houghton Mifflin.
Thoreau, H. D. (1973). "Resistance to civil government." In *The writings of Henry David Thoreau: The higher law*. Princeton: Princeton University Press.
Tocqueville, A. (2001). *Democracy in America*. New York: Signet Classics.
Treanor, B. (2008). Narrative environmental virtue ethics: Phronesis without a phronimos. *Environmental Ethics*, 30(4), 361–380.
Wensveen, L. Van (2000). *Dirty virtues: The emergence of ecological virtue ethics*. Humanity Books.
Wensveen, L. Van (2005). The emergence of ecological virtue language. In R. Sandler & P. Cafaro (Eds.), *Environmental virtue ethics*. New York: Rowman and Littlefield.

J Agric Environ Ethics (2010) 23:29–42
DOI 10.1007/s10806-009-9182-5

ARTICLES

Forgivingness, Pessimism, and Environmental Citizenship

Kathryn J. Norlock

Accepted: 27 May 2009 / Published online: 12 June 2009

Abstract Our attitudes toward human culpability for environmental problems have moral and emotional import, influencing our basic capacities for believing cooperative action and environmental repair are even possible. In this paper, I suggest that having the virtue of forgivingness as a response to environmental harm is generally good for moral character, preserving us from morally risky varieties of pessimism and despair. I define forgivingness as a forward-looking disposition based on Robin Dillon's conception of preservative forgiveness, a preparation to be deeply and abidingly accepting yet expecting human error. As with other virtues, however, preservative forgiveness is available to some of us more than others; in the second half of this paper, I consider the deep challenge posed by rational pessimism, especially on the part of those who have been given many reasons not to hope for the very moral improvements for which they strive. I conclude that for those of us with the power roles and personal resources especially conducive to environmental activism, preservative forgiveness inclines us to remain engaged in environmental activism with fellow flawed human beings, recognizing our own mutual depredations while committing us to cooperatively respond.

Keywords Forgiveness · Virtue · Pessimism · Ecocitizenship

Introduction

Philosophers of forgiveness disagree as to whether forgiveness is, or should be, considered a virtue. I am not usually one of the philosophers who argue that forgiveness ought to be understood primarily as a virtue. However, in what follows I argue for the cultivation of a virtue of forgiveness, specifically for a variety of

K. J. Norlock (✉)
St. Mary's College of Maryland, 18952 E. Fisher Rd., St. Mary's City, MD 20686, USA
e-mail: kjnorlock@smcm.edu

preservative forgiveness, that is, a predisposition to be forgiving in light of the inevitability of human error. In the context of environmental ethics, in particular, I presume it is uncontroversial to state that culpable human actions have been, and will continue to be, sources of environmental degradation. Our attitudes toward human culpability for environmental problems have moral and emotional import, influencing our basic capacities for believing cooperative action and environmental repair are even possible.

In short, I am arguing that preservative forgiveness promotes good *ecocitizenship*, and specifically, those practices of ecocitizenship that take the form of environmental activism and dedication to policy change. I use the term in a narrower sense than its originator intended, but in a way compatible with its broad definition. Broadly, as Robert Hull characterizes it, ecocitizenship "is not the whole of human excellence," but constitutive of it, and includes "the recognition that we are all plain citizens of our planet, that we for our own physical, intellectual, and moral benefit share it with other forms of life," and that fellow human beings are among those with whom we must share the planet well (Hull 2005). In the more narrow and literal sense of ecocitizenship, I argue that preservative forgiveness promotes one's *social and political engagement* with fellow human beings one would otherwise condemn as blameworthy for environmental depredations, which can, when taken to extremes, result in the vice Hull identifies as ecomisanthropy. Hull notes that ecomisanthropy, "the tendency to undervalue grossly or completely ignore human interests in favor of wild nature, is rare," but I am concerned about intermediate states short of misanthropy, states in which one's rational pessimism that humans are capable of or likely to change their environmental practices promote one's despair, disengagement, and consequently, inaction (Hull 2005). I suggest that the environmental crises we currently face are so significant that concerted collective action and changes in environmental policy are indeed requisite to a life of human flourishing. Therefore, dispositions that carry the risk of disengagement, despair, and inaction are dispositions that an environmental virtue guards against. It is the project for the rest of this paper to show that preservative forgiveness assists in reducing such risks.

Having the virtue of forgivingness as a response to environmental harm is generally good for moral character, preventing excesses of excusing and condemning, so that we locate responsibilities rather than sliding into attitudes of helplessness or recrimination, which foster inaction or misdirect our energies. Further, cultivating forgivingness is especially urgent for those of us who enjoy positions of power that require public and ongoing responsiveness to the environmental depredations of fellow human beings. As with other virtues, however, preservative forgiveness is available to some of us more than others; in the second half of this paper, I consider the deep challenge posed by rational pessimism, especially on the part of those who have been given many reasons not to hope for the very moral improvements for which they strive. I conclude that for those of us with the power roles and personal resources especially conducive to environmental activism, preservative forgiveness has a role to play in our environmental ethical lives.

Part One: Forgiveness as a Virtue

What it means to say that anything is a virtue bears some explanation, and the virtue of forgiveness, in particular, is one that requires some construction; as philosophers have noted, forgiveness is not a central virtue for the ancient Greek philosophers from whom we derive our theoretical frameworks. My own working definition of virtues relies on an Aristotelian framework; I characterize virtues as the appropriate states and dispositions to act of an excellent person. When one acts in accord with the virtues, one finds the contextually appropriate mean between two extremes. Virtues are not universal rules but the excellent practices of desirable and essential character traits that promote our realization of our natures and our happiness, so not everything that admits of a mean between extremes can be a virtue. For instance, there is no virtue of lying, even though everyday moral life might allow for an excess of lying (mendacity) and a deficiency of lying (such as brutal honesty regardless of circumstances).

Charles Griswold offers the valuable reminder that Aristotle's is a "perfectionist ethical scheme, for it is one that seeks to articulate and recommend the character of the man—and in Aristotle, it is a man—of complete virtue," a self-sufficient and magnanimous man whose relationships, true friendships, are with similarly virtuous characters (Griswold 2007). The virtue of forgiveness, therefore, does not appear central to the life of an excellent person, since forgiveness presupposes the wrongdoing of imperfect others and one's own vulnerability to resentment and vengefulness. With Griswold and others, I prefer to steer away from perfectionist depictions of agents as "immune from receiving injury," and proceed from a view of humans as characterized by fallibility; I take it to be fundamentally morally important that we are subject to luck, that suffering is pervasive, and that we are "embodied, affective, and vulnerable creatures," which Griswold rightly notes changes "what will count as a virtue" (Griswold 2007). Margaret Urban Walker is persuasive that as vulnerable and emotional creatures, for whom relationships are vital, we require trust and hope in others for our well-being, perhaps our very persistence; an inclination toward forgiveness may be a precondition of relationships we can rely on to continue, especially if an excellent and happy life includes not just minimally decent relationships but some enduring and trust-filled relations as well, in which one can expect occasional forgiveness (Walker 2006).

If forgiveness is to be considered a virtue, then it must be possible to identify extremes of excess and deficiency. Philosophers who identify these extremes more often than not cast them in terms of their frequency over time, especially when describing the deficiency of forgivingness in a character. One who *never* forgives is usually described as petty, mean-spirited, resentful, begrudging, hardhearted; vicious resentfulness can be seen as self-indulgent of one's hurts to an extent that inappropriately harms others. Interestingly, the extreme of excess comes in for much more debate among philosophers who are usually united in the view that forgiveness admits of deficiency. The nature of excessive forgiveness is sometimes similarly described as a matter of frequency, as philosophers debate the Christian commandment to forgive and the adage that to understand all is to forgive all; most theorists are no more comfortable with a frequency of one hundred percent than they are with zero (Hampton 1988; Murphy 2002). Yet the excess tends to come in for more complex treatment; its dimensions have variously been described as a ready tendency to be lenient toward offenders, to recommit to relationships no matter their worth, to extend oneself to accept wrongdoers at risk to one's safety or self-respect, or to condone culpable harm, to be hasty, to overlook, to be "other-indulgent" in the extreme at one's own expense.

Most of these failures involve the moral agent's inattention to context, to moral circumstances that require more respect, deliberation, or time than the forgiver has permitted. Many philosophers note the parallel arguments in Aristotelian and Nietzschean value theory that to bear any insult is slavish, but most forswear a prone slavishness as a vice of excessive forgiveness. Forgiving at least implies that one is cognizant of a harm that occasions enough bad feeling to call for forgiveness, and the doormat does not seem to exhibit even this much. It seems more accurate to describe being "too forgiving" as the disposition to be overly disposed regardless of the particulars—experiencing anger and acting in the right way, at the right time, toward the right persons, with "essential good judgment"—or too ready to rush through the processes that recovering from harms may require (Griswold 2007). Virtues require knowledge of the particulars to be exercised appropriately in different situations, and an excessively forgiving person may be inclined to proceed as though all harms are minor or trivial, or involve less than fully responsible agents.

To this emerging picture of the virtue of forgiveness, I wish to add a further distinction. Robin Dillon's excellent work on self-forgiveness includes the notion of "preservative self-forgiveness," as opposed to forgiveness for past incidents that requires transforming one's feelings about an offense (Dillon 2001). While the latter sort of forgiveness is backward-looking, pertaining to recovery from past harms, preservative forgiveness is forward-looking, a preparation to be deeply and abidingly accepting, with three features. The first is a view to humans "both optimistic and realistic," as "basically decent and aiming to be good, capable of moral self-improvement and of regaining their bearings if they get off track, and on the other hand, as inherently fallible and liable to get even far off track." The second is a flexibility in one's standards, an avoidance of being "dominated by standards [persons] must not fail to meet," compatible with lofty ideals and strong principles. Third, such a forgiving person has qualities of patience and tolerance which are compatible with anger, but which indicate "an ability to see their faults and wrongs and still value" persons (Dillon 2001). Dillon likens preservative forgiveness to the sense in which we say a surface is forgiving, so that a dropped glass may bounce instead of shatter (Dillon 2001). In light of Dillon's metaphor, I shall be using the term *forgivingness* interchangeably with preservative forgiveness.

It may seem that a forward-looking disposition does not fit neatly into the framework of a virtue, at first. Preservative forgiveness does not wait for knowledge of particulars to inform one's attitude toward what others may do. However, such forgivingness has an instructive parallel in Aristotle's account of the mean between the two extremes of angry behavior, which he refers to as a virtue with no name. Aristotle suggests that the virtuous person "seems to err more in the direction of deficiency, since the mild [virtuous] person is ready to pardon, not eager to exact a penalty" (Aristotle 1999, 1126a1). Philosophers generally agree that forgiveness is compatible with punishment and distinct from pardon, so a disposition to pardon is not identical to a disposition to be forgiving. Yet, just as the virtuous person is inclined to pardon prior to knowing particulars, the forgiving person may be one inclined to reaccept and recommit to others prior to their offenses. Consider it a commitment to maintaining relationships, and a particular method of locating responsibility, punishing, or rectifying wrong, a method that presupposes the fallibility, the worth and the capabilities of offenders with whom one is prepared to continue to live. The latter may sound too much like reconciliation, which many theorists prefer to distinguish from forgiveness, yet in the context of environmental ethics, I suggest that we should take reconciliation to be an inevitable outcome of forgivingness; until we have the choice to live elsewhere, we must be reconciled to living with fellow residents of the planet.

Going about moral assessment with a predisposition to reconcile is undoubtedly familiar to parents and educators, who hold the well-being and the ongoing trust of particular individuals in their hands, and who presumably commit to their trusting dependents over long terms that exceed the occasions on which offspring and students may culpably err. It is less obvious whom we have in mind when we commit to preservative forgiveness in environmental ethics and politics; whom are we preparing to forgive? Environmental harms are not always discreet and easily identified, with beginnings and end-points; locating culpability for wrongdoing is further complicated because environmental harm is often the result of aggregate effects of the actions of sometimes random, sometimes formal collectives. If culpability entails guilty knowledge, then even when identifiable discreet harms can be traced to individuals, it is not always clear that those individuals know what they are doing, or fully appreciate the consequences of their acts. Ignorance and deep-seated denial are often considered candidates for excuse rather than forgiveness. Last, environmental harms are global; the actions of some can affect the lives of people on the other side of the world, and one may never know the sources of one's diminished well-being, so the trustee relationship has no obvious parallel in environmental relationships.

To clarify my view in light of such complexities, I return to the attitude Dillon described, the view "both optimistic and realistic," of humans as "basically decent and aiming to be good, capable of moral self-improvement," and, I would add, capable of moral repair and change in behavior. This need not be a view of humans as unlikely to do wrong; it does not entail thinking, for example, that people would never buy bottled water, but it does hold that people are capable of increasing their awareness and decreasing their purchases of bottled water—as indeed they have (Martin 2008). Preservative forgiveness is a possibility for anyone with the ability to deliberate about their attitude toward the worth of those human beings who are and will be responsible for environmental harms. We need not know the extent of individuals' blameworthiness in order to reflect on how we are disposed to behold responsible persons (depending on the form of responsibility). Humans are likely to be culpable for future harms, and it need not be the case that each member of humanity is equally to blame in order for us to generalize about our attitudes toward humans' errors. Peter French argues that rather than focus on the extent of each member's responsibility within a collective, we should attend to the power the members have in their group to be able to move a group to action or inaction (French 1992). That is, instead of demonstrating a lack of responsibility for past harms, group members should demonstrate a lack of power to rectify or call attention to those harms and prevent future ones. Similarly, Margaret Walker urges our attention toward questions of moral repair, capacities for responsiveness, rather than questions of precision in blameworthiness (Walker 2006). Therefore, we can better account for reported attitudes in journalistic accounts toward whole groups and whole nations. One may sensibly say that one is disposed to hold particular attitudes toward "America" for heavy reliance on fossil-fuels, or toward "China" for increased sport-utility vehicle purchases. We can and do possess moral attitudes with respect to the responsibilities of groups and of humanity in general.

Since attitudes toward groups or humanity can be morally wrong or consequentially bad, it is perhaps an attractive option to say we should not cultivate any moral dispositions to see whole groups or all humans as culpably environmentally harmful, but I suggest this is not the route to take. Just as Strawson says it is neither rational nor desirable that we can renounce participant reactive attitudes, I would suggest it is neither productive nor possible to avoid having any beliefs or feelings about humankind's propensity to harm the environment, although one may hold these unreflectively (Strawson 1993). When one's beliefs lend one to certain attitudes and practices, such as withdrawing from environmental repair efforts or engaging in indiscriminate condemnation, then one may exhibit vices of incivility, apathy, or hopelessness. For those of us who have become or feel in danger of becoming such withdrawn or disengaged agents, preservative forgiveness, which takes the view that humans are both fallible and capable, is to be recommended insofar as it lends us to practices of continual recommitment to each other, in order to live with each other as we locate responsibilities and, more importantly, identify the powers that groups and members have to move each other to improve environmental practices. When beliefs about human culpability do not lend one to withdrawal and disengagement, preservative forgiveness may be unnecessary or out of reach and not worth further personal energy. One with a jaundiced view of humanity who still continues to plug away may persist without forgivingness; the rest of us, however, may find that without cultivated commitment to relationships with humanity, what begins as self-protective preparation to be disappointed slides into hopelessness and surrender. I am not the first philosopher to observe that disgust with my fellow human beings can incline me to apathy; close on the heels of the thought that someone is not worth saving is the thought that I need not bother.

In short, dispositions to blame come with personal and moral risks, although I grant that personalities vary and some of us end up at greater risk than others. Because surrender and inaction are outcomes both possible and worth avoiding, I suggest that preservative forgiveness is a virtue worth cultivating in anyone capable of acting on attributions of blame for environmental harm. The question as to who has standing to forgive is much more vexing in discussions of incident-forgiveness, in which philosophers debate who counts as the direct victim of particular discreet wrongs. Preservative forgiveness, in contrast, is an option for anyone who can reflect on what attitude to take towards fellow fallible persons (although some of us have this option more readily available than others). Virtues may be more incumbent on us in some domains than in others; those in roles that require them to establish or publicly evaluate environmental policies are predictably tested more often in their abilities to be deeply and abidingly accepting of the worth and potential of other persons, and to see individuals as more than the sums of their acts (Badhwar 1996).

If preservative forgiveness is practiced virtuously, then it must be exercised toward the right people and at the right times. Yet some are inappropriately approached with an attitude of forgiveness; if they are victims of environmental harm, or causally involved but powerless to do otherwise, there is nothing to forgive because there is no culpable wrongdoing. Still others may arguably be undeserving of forgiveness, for example, if they unrepentantly and knowingly pursue environmentally harmful practices because they are highly motivated by personal monetary gain. If forgiveness of victims or criminals is arrogant or hasty, then one might object that other virtues would seem to be more critical to environmental civil life, such as humility or a vigilant sense of justice.

I do not wish to deny that victims and wanton perpetrators of environmental harms exist; there certainly seem to be more victims than not, if indeed half of all humans live in poverty and therefore have few choices as to how to live decently (United Nations Human Development Report 2007/2008). For constrained victims of dire harm, sympathy and care are more appropriate dispositions to motivate aid; for reckless, overtly evil perpetrators, condemnation and attention to their punishment may be obviously attractive alternatives. However, I suggest that preservative forgiveness is the appropriate attitude to take to many, even most, fellow humans, that is, to those who are neither moral innocents nor moral monsters as they conduct their environmental lives. Most Americans, for instance, live in suburbs and continue to drive cars on a daily basis, a development neither entirely of their making nor entirely unchosen. Most American households own computers now, and academics fire them up to write about the recent surge in our consumption of fossil fuels; offices boot up across the country every morning, such a staple of the workplace that its environmental impact is obscured by its ubiquity. Unlike incident-forgiveness for past acts, preservative forgiveness does not entail attributions of full moral condemnation to individuals for particular acts. Instead, preservative forgiveness expects that those of us faced with choices are likely to err, as Dillon says, to go even far off track, for reasons that are mixed, even mysterious, generating consequences that are sometimes transparently foreseeable, sometimes not.

Yet certainly, the power and responsibility of different individuals may vary depending on context. Those who enjoy reasonably secure lives and enough wealth to permit choices of lifestyle may be held to higher standards of environmental awareness and responsibility, though here, again, comfortable people will regularly fail to meet those standards. If anything, knowledge of their comfort and wealth may incline us to be unforgiving. In a frustrating reversal of the saying, "To understand all is to forgive all," many of us who live in industrialized and developed nations know quite well the details of our robust environmental options and consequent culpabilities; while we may excuse those far away whose sufferings and constraints we do not fully know or share, knowledge of particulars close to home tax our dispositions to forgivingness of fellow citizens. Again, the virtue may be called for in some domains more than others. Because the very preconditions of exercising virtue, such as knowledge of particulars and opportunities to act, may also make the virtue of forgivingness more difficult to live up to, I suggest that preservative forgiveness is especially critical in our domestic political and civil lives, that is, in those contexts in which we know all too well what we must do collectively and how we fail to do it. In my conclusion, I outline evidence for forgivingness as conducive to social engagement, which I suggest is most pressingly necessary at those times when we'd rather do the opposite of engage, when we'd rather give up and leave citizens to their fates. As Herman Melville says, at those times when our dispositions toward our neighbors is most taxed, we'd sooner do the opposite of social engagement; we'd rather hang it all and take to the sea than reconnect with people, and at such times "it requires a strong moral principle to prevent me from deliberately stepping into the street, and methodically knocking people's hats off" (Melville 2008). Yet abandoning one's community does not contribute to solving environmental problems requiring collective action. In the next section, I consider alternatives that lie between forgivingness and moral exit.

Part Two: Considering Counterarguments for Rational Pessimism

To environmental philosophers, preservative forgiveness may not seem called for, especially given Dillon's recommendation to be flexible in one's standards. Concerned and observant people may see the U.S. as suffering from deficiencies of environmental justice rather than forgiveness. To many, forgiveness further connotes a reluctance to punish—note that one with Aristotle's virtue of mildness is not "eager to exact a penalty;" it is reasonable to object that any disposition that promotes ignoring past harm is undesirable. Environmentally harmful practitioners are scarcely held accountable often enough as it is, one could point out, and so a more pessimistic disposition to expect offenders to be unrepentant, and to hold wrongdoers to their acts, is strategically superior to an optimistic view of humans as willing to change and worth continual reacceptance and recommitment. Although a more pessimistic disposition does not preclude future occasions for incident-forgiveness, one with a predisposition to be forgiving would seem overly inclined to think better of environmental wrongdoers than they deserve, or better than past behaviors would give good grounds for holding.

A pessimistic view need not be cynical, as Lynne McFall demonstrates in her defense of bitterness. While acknowledging that bitterness can be, and generally is, harmful to oneself, McFall argues that there are circumstances in which "active bitterness is not always bad for its host" (McFall 1991). Bitterness may be rational and justified, according to McFall, and in its active form can include "intense animosity or virulence of feeling;" these are appropriate as responses to the culpable harms inflicted by human agents from whom one had legitimate hopes of better treatment (McFall 1991). Similar to unforgiving anger, bitterness exceeds merely being unforgiving by seeking retribution, even by being vindictive, but McFall suggests this is a justified response when one must actively "move away from self-deception" toward such responses as "truth-telling and bearing witness;" in some cases, one ought to want vindication (McFall 1991). As Susan Brison points out, bearing witness is an optimistic act, entailing the belief that one is or will be in a position to be heard (Brison 2002). Bitterness rebukes, and rebuke entails an audience.

So described, bitterness may not be an optimistic disposition but a pessimistic disposition that allows room for optimistic acts and hope for change. Bitterness declares one's worth and asserts an implicit demand for something better. With its source in past disappointments and harms, bitterness may be both unavoidable and rational for those in the worst of circumstances, but because it sometimes promotes justice-seeking actions, McFall defends it as preferable to cynicism. Cynicism involves seeing one's hopes for better treatment as false; the adage that "life is unfair" is McFall's example of the cynical position (McFall 1991). Like preservative forgivers, cynics take the view that humans are fallible and likely to err, but cynics would disagree that most are also basically good and capable of (or likely to) change. Because cynics accept the unfairness of life and take the inevitability of human error for granted without hope of better, they effectively let environmental wrongdoers out of responsibility to rectify harms. Cynics cannot build a better nation if it is one in which we cannot expect better of citizens than unceasing environmental destruction; cynicism amounts to an environmental vice, failing to promoting environmental values and human flourishing. Seen this way, bitterness may hold out more hope for the future, since in its active form it could include correcting the course of a society and reminding us to learn from environmental evils. I will conclude with some reasons to prefer preservative forgiveness to bitterness, but before I do so, a more thoroughgoing and complicated variety of pessimism must be considered.

Although bitterness may be pessimistic in outlook but allow for optimistic acts, one may also have good reason to be pessimistic and hold *no* hope of realizing the changes one seeks. Lisa Tessman eloquently describes the social and political activism of pessimists with no hope of attaining that for which they struggle, adding, "If one sets one's sights high—on something highly desirable but highly improbable—one has little reason to expect success in attaining the object of one's striving," that is, little reason to predict that achieving one's goals is likely (Tessman 2009). Tessman does not specifically discuss environmental activism; I extend her concern for the well-being and persistence of particular oppressed people, who maintain a claim to flourishing that they have no hope of realizing for themselves, to the dispositions of those who aim for ambitious environmental policies that can realistically be achieved only after their own lifetimes are over, if at all. Note that the environmental goals need not be foolishly lofty to be unattainable; on the contrary, if, say, progress toward irreversible effects of global warming is too far along, then even aspirations for minimally requisite reduction of carbon emissions within a constrained time frame may be both highly desirable and highly improbable. Environmental activists are not all oppressed persons (although many victims of oppression are also environmental activists), yet I recognize many activist friends in Tessman's description of "those whose expectation for success (and perhaps also whose hope) is vanishingly small because their sights are set on the high aim of radically transforming oppressive societies and they realistically accept the heavy knowledge of the improbability of attaining this aim" (Tessman 2009).

So far, I have considered only *predictive expectations*, that is, predictions inferred from evidence, probability, or expressed intentions (Walker 2006; Tessman 2009). Tessman focuses on the loss involved in having neither predictive expectations nor *normative expectations*, that is, "expectations that people will act as they *should*" and that "embody a sense of *entitlement* to what we expect, and not merely an anticipation of it," even absent predictive expectation (Walker 2006; Tessman 2009). Tessman argues against Margaret Urban Walker's observation that when we have no predictive expectation, we still must have hope that normative expectations will be met, informed by basic trust in fellow human beings, "the real *possibility* (non-zero probability, less than certainty) of what is hoped for, in the estimate of the one who hopes" (Walker 2006). As Tessman argues persuasively, "sometimes one can realistically predict that there is essentially *no* possibility of success" (Tessman 2009). Yet sometimes cynics with neither hope nor trust in fellow humans persist in activism, and it is these activists Tessman and I seek to understand; their continued resistance give me reason to consider that my arguments for forgivingness are arguments for something quite possibly consequentially good but ultimately unnecessary.

The above constitute compelling reasons to reject preservative forgiveness toward fellow environmental actors. First, arguments that bitterness is rational and compatible with optimistic acts suggest that, in the absence of reasonable expectations of reciprocal commitment or acknowledgment of harm, publicizing rebuke would go further toward locating responsibility for environmental harm, and move third parties to punish those actors who persist in wrongdoing. As an especially vigilant variety of anger, bitterness may be most appropriate for those for whom others fail to stand up, asserting one's own outrage at injustice in a sustained and persistent way in order to assert one's own worth in the face of wider silence. For example, considering the linked environmental and social injustices suffered by residents of New Orleans in the aftermath of Hurricane Katrina, it would be unreasonable to urge residents to take optimistic views of the likelihood of excellent governmental responses to their losses.

Second, accurate predictive expectations that needed environmental goals cannot be achieved may be too compelling to comfortably coexist with optimistic views of humans as capable of change. More precisely, negative predictive expectations suggest it is more rational to see humans as capable of personal and behavioral change, even basically good, but outmatched by the seriousness of the problems we face. Consider the concern voiced in some quarters that melting polar ice may have already reached a point of no return. If so, then once again pessimism seems rationally called for to an extent that seemingly overshadows the usefulness of preservative forgiveness, and one is left with a retreat to "developing adaptive preferences, that is, preferences for what is within the range of options that are available," such as goals to reduce carbon emissions slightly and over the course of a decade or two, making a bad situation more livable rather than striving for highly desirable, flourishing states (Tessman 2009).

Arguments for the rationality of pessimism constitute serious objections to the appropriateness of preservative forgiveness toward fellow environmental actors, for some people. Yet to return to the framework of virtue, Tessman persuades me that the appropriateness of pessimism varies with the circumstances of moral agents, specifically with their positions of power and their personal resources, and in the case of the varieties of environmental activism with which I'm concerned, with the obstacles they face or the advantages they have in asserting their social and political power to move others to collective action. Although victims of serious harm and oppression can have an overabundance of evidence that pessimism is warranted, less constrained agents have reason to believe they are well positioned to be effective in bringing about the conditions that make some environmental goals attainable. Many Americans have the money, leisure, and political freedoms necessary to reduce their own energy consumption; some Americans have considerable influence and opportunities to persuade others to do the same. Since the United States is by far the greatest contributor to global environmental problems, the choices of its most privileged citizens would make a great deal of difference in what it means to say particular environmental goals are attainable. Therefore, accurate negative predictions are contingent upon our behaviors, changes that further depend upon our beliefs about whether or not changing our behaviors will be efficacious. Since behavioral change will only be efficacious *en masse*, it is incumbent upon the very citizens who enjoy most of the benefits of a wealthy country and attendant high energy consumption to cultivate a disposition to see themselves and their fellow citizens as capable and willing to change environmental behaviors. If we do not believe we are capable of collective action, we will have no motivating expectation to act. And indeed, the more advantages we each have individually, the more reason we have to think that we are capable of influencing each other and affecting policy. Some of us are more capable than others.

Aristotle notes that a life of good fortune is supportive of living virtuously, and that serial and serious sufferings tax our ability to live according to certain virtues (Aristotle 1999, 1100a8–9, 1101a6, 11–12). Although happiness is not identical with prosperity, Aristotle points out that a happy and virtuous life consists in activities that require some external goods in order for it to be possible that those activities can be exercised (Aristotle 1999, 1099b7–8). I prefer, for the current purposes, to argue the obligations of virtue also run in the other direction: Given the preconditions for virtue found in a blessed and fortunate life, we can rightly hold lucky people to the expectation that they cultivate dispositions that will incline themselves and their fellows to excellent environmental citizenship, that is, to forgivingness and its optimistic view of humans as both fallible and capable of, even eager to, be better environmental actors.

Conclusion: Preservative Forgiveness for Self-preservation

In light of the argument above that bitterness is appropriate when it promotes rebuke and resists silence, I suggest that those who do *not* suffer in silence, those with rather more powerful voices, whom it is reasonable to expect will have a receptive audience, act virtuously when they avoid bitterness and instead cultivate optimism. When one *can* expect the attention of fellow citizens, one ought to exercise well-founded normative expectations of others' engagement. On such an argument, the most powerful members of our society are doing exactly what they ought, and not acting supererogatorily, when, for instance, Vice President Al Gore raises awareness of global warming.

Further, empirical evidence increasingly suggests that moral agents who cultivate forgiving dispositions are more likely to persist in prosocial behavior, by which I mean a pattern of voluntary helping behaviors in interpersonal and collective settings, a willingness to contribute to social or political goals beyond one's required role; prosocial behavior implicitly involves seeing the relationships involved in such behaviors as having value (Eisenberg and Mussen 1989; Bowler and Brass 2006). Now, contribution to collective effort is observable of many agents who are not forgiving, and I am not arguing in the scope of this paper that prosocial behavior requires a forgiving disposition. Indeed, I do not believe forgivingness is a necessary condition of all engaged and helping behaviors. I am instead pointing to the positive evidence that at those times, and in advance of those times, when collectives and groups experience conflict, agents with dispositional forgivingness typically cope skillfully with disappointment, relational setbacks, and the internal stresses involved in conflict; a tendency to be forgiving "typically serves as a resilience factor that promotes well-being," and the maintenance of relationships over the course of one's contribution to collective effort (Ysseldyk and Matheson 2008; Strelan and Covic 2006; Worthington 2006; Snyder 2001). Forgivingness is both interpersonally and intrapersonally beneficial, and it is the latter that specifically interests me (Denton and Martin 1998). My concern throughout this essay has been with the internal disposition to have low expectations of others, to be pessimistic to the point of reasonably considering that one should not bother acting at all. Given the focus of my attention, it is compelling to me that research shows those with "highly forgiving dispositions" are "inclined to endorse coping strategies that are typically considered to be adaptive, such as problem solving and cognitive restructuring," that is, looking for what one has learned from the situation in order to continue better contributing to the group (Ysseldyk and Matheson 2008). Forgiving dispositions also indicate, in multiple studies, "a reduced propensity to …rumination and wishful thinking," which tend to be maladaptive, resulting in disengagement with others and cessation of efforts, especially efforts in the direction of prosocial activity (Ysseldyk and Matheson 2008).

In arguing for the benefits of forgivingness to intrapersonal well-being, I hope I have shown that reduced stress, despair, and contempt are not merely self-regarding, but also inherently tied to one's social conduct and one's motivation to participate in collective ends. As Aristotle reminds us, in doing fine actions, the virtuous person benefits himself while benefiting his friends. To be a good citizen is also to be an excellent person; dispositions that lend the moral agent to be happy and virtuous also promote one's proper conduct in the polis. Therefore, it is consistent with virtue to be forgiving for one's own sake every bit as much as for others. Consider the contrast to the view, widely held among philosophers of incident-forgiveness, that forgiveness is not appropriate when done to make one feel better or to achieve merely consequential aims. Aristotle's account of virtue gives us reason to believe that dispositions and practices like forgivingness that make us happy and feel good (or, at least, less bad in some sense), while relieving political conflict and promoting civil discourse, are proper to excellent lives and virtuous characters. Friends may be the best external good and membership in a polis may be part of an excellent life, but one must also live with oneself. Cultivating preservative forgiveness toward others may ease one's own internal stresses and make it easier to communicate with those one would otherwise condemn or cease engaging with in political coalition.

Forgivingness would seem to be virtuous even in the presence of negative predictive expectations. The optimistic normative expectation that humans will be forgivable is not dependent upon what we can predict about our actual chances for adequate environmental repair. After all, a virtue ethic is not sheerly consequentialist; one does not pursue virtue if and only if good policies really will result from one's actions. One lives a virtuous life for its own sake, in order to be happy and to flourish as one's own activities promote one's capacities to do so. Were we to discover that ambitious policies will not entirely succeed, we would be in something like the position Aristotle describes, of the man of virtue who comes upon misfortune; our abilities to pursue excellent environmental activities would be tested, ultimately impeded, but "even here what is fine shines through," not because we feel no distress, but because we are constant in our commitments to living well and working with others in pursuit of desirable environmental goals (Aristotle 1999, 1100b31).

My account of forgivingness presupposes that environmental actions are collaborative, and that our most pressing environmental problems have social and political valences. For environmental activities that can be pursued in isolation, forgivingness is arguably not necessary. One could pick up an aluminum can and drop it in a recycling container with a sort of glowing Kantian sense that one is doing one's duty regardless of whether or not anyone else recycles; indeed, on some (unconvincing) interpretations, recycling the can is most transparently morally worthy when one doesn't even want to recycle, let alone feeling motivated by optimistic expectations of others' recycling. Yet I wish to close with the pragmatic argument that it is not enough to practice solitary environmental acts from a good will that shines like a jewel in its own right. To flourish and to believe future generations will flourish is part of a happy life; a teleological virtue ethic makes sense of the inherent importance of environmental citizenship to individual happiness. Especially in countries that consume the most energy and produce the most waste, individuals and communities must come together to respond to massive environmental problems, which requires some willingness to believe in ourselves and in each other. Forgivingness, like Aristotle's virtue of mildness, inclines us to act, recognizing our own mutual depredations while committing us to cooperatively respond.

References

Aristotle. (1999). *Nicomachean ethics.* (trans: T. Irwin). 2nd ed., Indianapolis/Cambridge: Hackett Publishing Company.

Badhwar, N. K. (1996). The limited unity of virtue. *Noûs, 30*(3), 306–329.

Bowler, M., & Brass, D. J. (2006). Relational correlates of interpersonal citizenship behavior: A social network perspective. *Journal of Applied Psychology, 91*(1), 70–82.

Brison, S. (2002). *Aftermath: Violence and the remaking of a self.* Princeton, NJ: Princeton University Press.

Denton, R. T., & Martin, M. W. (1998). Defining forgiveness: An empirical exploration of process and role. *American Journal of Family Therapy, 26*(4), 281–292.

Dillon, R. (2001). Self-forgiveness and self-respect. *Ethics, 112*, 53–83.

Eisenberg, N., & Mussen, P. H. (1989). *The roots of prosocial behavior in children.* Cambridge: Cambridge University Press.

French, P. (1992). *Responsibility matters.* Lawrence, KS: University Press of Kansas.

Griswold, C. (2007). *Forgiveness: A philosophical exploration.* Cambridge: Cambridge University Press.

Hampton, J. (1988). *Forgiveness and mercy.* Cambridge: Cambridge University Press.

Hull, R. (2005). All about eve: A report on environmental virtue ethics today. *Ethics and the Environment, 10*(1), 89–110.

Martin, A. (2008). Tap water's popularity forces pepsi to cut jobs. *New York Times*, October 14; Accessed online 8 November 2008.

McFall, L. (1991). What's wrong with bitterness? In C. Card (Ed.), *Feminist ethics* (pp. 146–160). Lawrence, KS: University Press of Kansas.

Melville, H. (2008). Moby Dick. In T. Tanner (Ed.), *Loomings* (Oxford World Classics edition). Oxford: Oxford University Press.

Murphy, J. (2002). Forgiveness in counseling: A philosophical perspective. In S. Lamb & J. Murphy (Eds.), *Before forgiving: Cautionary views of forgiveness in psychotherapy* (pp. 41–53). Oxford: Oxford University Press.

Snyder, C. R. (Ed.). (2001). *Coping with stress: Effective people and processes.* New York: Oxford University Press.

Strawson, P. (1993). Freedom and resentment. In J. M. Fischer & M. Ravizza (Eds.), *Perspectives on moral responsibility* (pp. 45–66). Ithaca: Cornell University Press.

Strelan, P., & Covic, T. (2006). A review of forgiveness process models and a coping framework to guide future research. *Journal of Social & Clinical Psychology, 25*(10), 1059–1085.

Tessman, L. (2009). Expecting Bad Luck. *Hypatia, 24*(1), 9–28.

United Nations Human Development Report. (2007/2008). New York: United Nations Development Programme. First Published in 2007 by Palgrave Macmillan.

Walker, M. U. (2006). *Moral repair: Reconstructing moral relations after wrongdoing.* Cambridge: Cambridge University Press.

Worthington, E. L., Jr. (2006). *Forgiveness and reconciliation: Theory and application.* New York: Routledge.

Ysseldyk, R., & Matheson, K. (2008). Forgiveness and coping. In W. Malcolm, N. DeCourville, & K. Belicki (Eds.), *Women's reflections on the complexities of forgiveness* (pp. 143–164). New York: Routledge.

J Agric Environ Ethics (2010) 23:43–59
DOI 10.1007/s10806-009-9185-2

ARTICLES

Radical Hope for Living Well in a Warmer World

Allen Thompson

Accepted: 27 May 2009 / Published online: 17 June 2009

Abstract Environmental changes can bear upon the environmental virtues, having effects not only on the conditions of their application but also altering the concepts themselves. I argue that impending radical changes in global climate will likely precipitate significant changes in the dominate world culture of consumerism and then consider how these changes could alter the moral landscape, particularly culturally thick conceptions of the environmental virtues. According to Jonathan Lear, as the last principal chief of the Crow Nation, Plenty Coups exhibited the virtue of "radical hope," a novel form of courage appropriate to a culture in crisis. I explore what radical hope may look like today, arguing how it should broadly affect our environmental character and that a framework for future environmental virtues will involve a diminished place for valuing naturalness as autonomy from human interference.

Keywords Climate change · Consumerism · Courage · Hope · Responsibility

Environmental changes bear upon the environmental virtues. For example, glaciers around the world are in retreat due to global warming, while a "majority of people in the world depend on water from glacial melt for food, drinking water, and irrigation" (Gartner 2008). From the perspective of the future, then, the way many people use water today might rightly be seen as excessive and thus the character of a person so disposed as profligate, perhaps ignoble, thoughtless, or even unjust. "What may not be greedy in other circumstances," Jason Kawall recently observed, "may be greedy given current global conditions" ("Reconceiving Greed in an Age of Global Climate Change," unpublished manuscript).

A. Thompson (✉)
Department of Philosophy and Religion, Clemson University, 126 D Hardin Hall, Clemson, SC 29634, USA
e-mail: athomp6@clemson.edu

Instead of identifying new *applications* of existing virtue terms, I'm concerned with the possibility that some of our virtue *concepts* may change. I argue that the impending radical changes in global climate will likely precipitate significant changes in the dominant world culture and then consider how these changes could alter the moral landscape, particularly culturally thick conceptions of the environmental virtues. Our understanding of environmental virtues may be moored to cultural and environmental pilings that are unstable and we should begin to consider the kinds of character traits best suited to radical change.

In the first section I describe how consumer culture is vulnerable and argue it's reasonable to expect that some of our contemporary ways of going on simply will not survive. In the second section I outline some ideas about virtue in the face of cultural devastation from Jonathan Lear's account of Plenty Coups, focusing on "radical hope," a novel form of courage appropriate to a culture in crisis (Lear 2006). Next, I explore what radical hope may look like today, arguing that it should broadly affect our environmental character and that a framework for future environmental virtues will involve a diminished place for valuing naturalness as autonomy from human interference. My virtue-theoretic orientation is broadly eudaimonistic, according to which the *environmental* virtues depend upon connecting what it is to live well with our basic axiological conceptions of the extra-human world. The view I try to make sensible here is that future generations will have to develop some new environmental virtues suited to a new world environment. This will include learning what it is to flourish while being responsible for the global biosphere through those manifold, ordinary, and everyday activities characteristic of our human form of life.

Consumerism and Cultural Change

Dominant cultural patterns of economically privileged societies worldwide have become more homogenous since the middle of the 20th century, converging on what has been called the consumer society. By "consumerism" I mean a set of attitudes and values leading people always to high levels of consumption and orienting them to find meaning and satisfaction in life largely through the practices of purchasing new consumer goods (Goodwin et al. 2007). Today's consumer culture would not be possible without the Industrial Revolution and so is intimately connected, at least historically, to the burning of fossil fuels for energy. Industrial society has enabled a culture of consumerism, while the habits, values, and standards actually cultivated in consumer societies affect the background assumptions, concepts, and practices of the people living in it. I'll take it for granted that much of the contemporary value and conceptual landscape regarding the non-human world, including some of our ideas about the environmental virtues, takes its shape as a result of living in an industrial society, socialization into a culture of consumerism, and a life of material wealth.

Wes Jackson offers four ideas that illustrate the vulnerability of our consumer culture and support the claim that the near future will likely bring significant change.[1] Together, Jackson's points reflect the anomaly of our times. When conditions that enable consumer culture are suspended, many of the forms of life adopted within it will no longer be of much use, perhaps amounting only to outdated habits, ill-suited to living well under a new world climate.

Jackson's first point: no person who died before 1930 lived to see the world's human population double and, experts predict, neither will any person born after 2050. This means that by 2050 only about 0.066% of all human generations will be those to have lived through a doubling of the global population (Dillard 1998). Unlike the vast majority of people who will ever live, we are accustomed to background base-rates of population growth accompanied by the astounding technological developments that are characteristic of our times and drive the economic growth that makes consumerism possible, a lifestyle that only a relatively small fraction of the world's current population enjoys. Yet in some basic ways what *we* take to be normal likely will be, in the full course of human history, an outstanding aberration. If this is right, then when we stop growing and return to some kind of equilibrium, we should expect also that many, but not all, of the culturally embedded perspectives and habits basic to late twentieth century life will not continue.

While exponential global human population growth is historically novel, the exploitation of carbon-based forms of energy is not. To the contrary, Jackson's second point: *Homo sapiens* have been exploiting carbon-based forms of energy throughout history, as relentlessly as existing technologies would permit. The birth of agriculture began a long history of harvesting rewards from soft carbon nutrients in soil, essentially a non-renewable resource. Next we harvested energy stored as carbon in wood, smelting ores in the Iron and Bronze ages, and today we burn the "condensed ancient sunlight" of fossil fuels—coal, oil, and natural gas—to drive our growth economy, enabling our modern lifestyles.

Thus, the challenge we face in preventing further global warming is to develop the first human society *ever*—and very quickly—that is not based on exploiting carbon-based forms of energy. There has *never been* such a society! If we are able to meet this challenge, then there will literally be a new form of human civilization and most likely new cultural forms to fit. Someone might object that an adequate source of alternative, clean energy would enable the preservation of our consumer lifestyle. I'll return to consider this objection but, for now, let's suppose that abundant supplies of clean energy will *not* arrive in time to avert the drastic changes that we can otherwise expect to the lifestyles typical of today's industrial and economically developed societies.

Another reason to believe that significant cultural changes may be imminent appears when we connect facts about human population growth *with* our use of carbon fossil energy, especially natural gas and oil. Jackson's third point: it is no coincidence that the sudden upward curve of human population closely parallels that of oil production, once we have the technology to turn fossilized carbon into

[1] From Jackson's presentation at *Ethical, Cultural, and Civic Dimensions of Global Climate Change* hosted by the Center for Humans and Nature in Charleston, SC, 28 November 2007. In personal communication Jackson attributed the first point to (Cohen 2005).

artificial, ammonia nitrate fertilizer (with the Harber-Bosch process, invented in 1913). Human population and the use of *some form* of carbon-based energy have been on a mutual growth trajectory for a long time. But fossil carbon, industrial technologies, and artificial fertilizer have put this trajectory on an accelerated arch. Once oil and other fossil fuels are gone—and many experts believe we are either now at or are quickly approaching the condition of peak oil—the human population must fall to a level that can be supported by an energy economy supplied by contemporary sunlight (solar and wind), supplemented by other sources such as tidal, geothermal, and nuclear. Without technology, the size of this population has never been more than one billion people. It strikes me as unlikely that first-world cultures during or after a significant global population crash would be very much like our own.

Jackson's forth point: over the course of his lifetime (b. 1936), we have burned 97% of all the oil that has ever been burned. By the end of my own lifetime (~d. 2050) we will have consumed close to *all* the oil on Earth. Without clean energy substitutes, how could future generations meet well the conditions of their world with only the practices, habits, and character dispositions we cultivate today, at the peak of this energy party? Almost everything characteristic of our modern way of life, from abundant food to the ubiquity of plastics, from cheap and abundant energy to apparently endless economic growth, depends upon the energy in hydrocarbons. From the long perspective of human history, it's not difficult to predict that this energy orgy will be relatively abrupt.

A culture based in modes of living that are utterly dependent on natural gas and oil is vulnerable to diminishing supplies. But let's suppose there is enough *coal* to continue powering our industrial, consumer society. There would be still another problem; burning fossil fuel is a principal cause of anthropogenic climate change. Several years ago James Hansen predicted that we have about 10 years to make drastic reductions in greenhouse gas emissions before passing unknown "tipping points" that will effectively make global warming unstoppable. Addressing the same point in terms of the atmospheric concentration of carbon dioxide, Hansen writes,

> "[i]f humanity wishes to preserve a planet similar to that on which civilization developed and to which life on Earth is adapted, paleoclimate evidence and ongoing climate change suggest that CO_2 will need to be reduced from its current 385 ppm to at most 350 ppm.... If the present overshoot of this target CO_2 is not brief, there is a possibility of seeding irreversible catastrophic effects" (Hansen et al. 2008).

I will not argue further for the claim that if we do not prevent the worst possible consequences of global climate change, then our contemporary, consumer culture will simply be impossible.

Now consider a very different outcome: international co-operation reducing annual global greenhouse gas (GHG) emissions worldwide to 80% of the 1990 level by the year 2050. If emissions were to peak at 2015, then to achieve this goal annual reductions would have to be *over* 4%. If emissions peak later, then annual reductions would be greater. In sum, without significant alternative sources of energy, even if we begin now, steep and persistent reductions will be the norm for about two generations.[2]

If we manage to stabilize atmospheric concentrations at 350 ppm, we may avoid further environmental catastrophe but would still undergo significant cultural changes. To see this, assume that GHG emissions correlate with the levels of energy used to support contemporary lifestyles. This means a 4% reduction in emissions correlates roughly with a 4% reduction in energy use and thus the economy, standards of living, and presumably one's perceived quality of life. Early reductions will come with relative ease, via increased conservation and efficiency. There will be a minimum effect on standards of living, in part because we will only be returning to the (high) levels of energy consumption characteristic the 1990s, plus we will have more efficient technologies and some forms of clean energy will come on-line. However, from the perspective of today's "standards of living," most of the required cuts will be tough, presumably requiring both lifestyles and social organization quite different from those that citizens of developed nations are accustomed to today. Imagine reducing energy consumption (the economy) 2–5% per year for forty consecutive years while carrying on as we do now, in terms of material consumption, *and* with an even greater percentage of the population doing it! Without significant amounts of alternative, clean energy this will simply not be possible.

The possibility of undergoing such radical cultural transition raises issues of human moral psychology. What if deeply ingrained ways of going on, practiced for generations, suddenly become inappropriate, simply impossible, or downright detrimental? How adaptive are we going to be? The best *short-term* response to the threat of global warming is political action realizing significant reductions in emissions and there is clearly an important role for virtues of environmental activism here, which I discuss below. What about the long-term? What might count as environmentally excellent traits of character in the culture of a post-oil economy, in a warmer and biologically less-rich world? A big part of the problem in answering this question is developing a vision of how to refit ourselves to a form of the good that outstrips the conceptual resources of contemporary culture. This kind of problem has been faced before.

Plenty Coups and Radical Hope

In *Radical Hope*, Jonathan Lear argues that Plenty Coups, last principal chief of the Crow Nation, lead his people *well* through a time of utter cultural devastation. According to Lear, when traditional Crow forms of life became impossible, including the opportunity to be courageous in battle, Plenty Coups' leadership exhibited the virtue of *radical hope* involving an *imaginative excellence*, which

[2] It may be the "greening" of our energy infrastructure, e.g., the production of huge numbers of photovoltaic panels or windmills, etc., will itself require an outlay of CO_2 emissions sufficient to trigger positive feedbacks and make global warming unstoppable. See (Monbiot 2008).

allowed his people to move forward courageously when the paradigmatic Crow forms of courage were no longer practically intelligible.

The Crow were nomadic hunters, a warrior culture developed over hundreds of years to a life of fierce competition with other tribes for control of a shifting territory and access to buffalo. In traditional Crow life, "*everything* counted either as hunting or fighting or as preparing to hunt and fight" (Lear 2006, 40). Unsurprisingly, they revered and celebrated the virtue of courage. Their paradigm of courage was the practice of planting a coup-stick, the primary use of which was to mark a boundary, a place past which no non-Crow shall be tolerated (Lear 2006, 23).

As a youth Plenty Coups was brought into the life of a Crow warrior through socialization and the internalization of particular ego-ideals. At that time, the Crow "were fighting to prevent *utter devastation* at the hands of the Sioux," the Lakota, and the Cheyenne (Lear 2006, 23). Were the Crow to have suffered a devastating holocaust at the hands of the Sioux, courageous leadership would have taken an easily identifiable form, i.e., the planting of a final coup stick. But in the later half of the nineteenth century, Crow survival was threatened more directly by European settlement. Once on the reservation, intertribal warfare was impossible and so was counting coup (Lear 2006, 31). When the Crow paradigm of courage was practically impossible, how could Plenty Coups lead courageously?

According to Lear, Plenty Coups's leadership was a courageous response to radical change. Under the new conditions, still emerging and unknown to the Crow, leading people through appropriate adaptation required outstanding exercise of practical wisdom (Lear 2006, 81). But acting wisely in the midst of utter cultural devastation required Plenty Coups to anticipate a future for the Crow that he did not yet know how to think about (Lear 2006, 32). What happened to the Crow was not just an unfortunate occurrence, "not even a devastating occurrence like a holocaust; it [was] a breakdown in the field in which occurrences occur" (Lear 2006, 34).[3] On Lear's interpretation, the Crow did not have the conceptual tools to grasp the changes that were underway nor, more importantly, how they could go on *as the Crow*. I am not suggesting that we are facing an "end of occurrences," but that impending environmental changes may spell the end for significant parts of our cultural perspective, including ways we are accustomed to conceiving of and valuing the natural environment and our received notions of responsibility (as I discuss below). It may now be difficult to understand how we can go on *as environmentalists*.

Plenty Coups was a celebrated warrior who had successfully internalized the ego-ideals of Crow society and led a life devoutly committed to realizing excellence as a warrior, of living courageously *as a Crow*. This makes facing the challenge of cultural crisis a problem of moral psychology.[4] Through a course of education and habituation a person's character takes shape, one's outlook becomes deeply

[3] This is how Lear interprets Plenty Coups's declaration that "[w]hen the buffalo went away the hearts of my people fell to the ground and they could not life them up again. *After this, nothing happened*" (Lear 2006, 2, italics added).

[4] "If, roughly speaking, we believe that *ought* implies *can*: if we think that in these challenging times people ought to find new ways—not just of surviving—but of living *well*, we need to give an account of how it might be psychologically possible to do so" (Lear 2006, 64).

ingrained and psychologically stable. This is especially true for the virtuous person, whose view of what is excellent, noble, and fine has become a psychologically "ingrained nexus of perception and motivation" (Lear 2006, 63). How could it be psychologically possible for the courageous Crow warrior, Plenty Coups, to alter his culturally thick conception of what it is to be courageous?

In response, Lear suggests that there might be "a certain plasticity deeply embedded in a culture's thick conception of courage" and defends the view that there may be "ways in which a person brought up in a culture's traditional understanding of courage might draw upon his own inner resources to broaden his understanding of what courage might be. In such a case, one would begin with a culture's thick understanding of courage; but one would somehow find ways to *thin it out*: find ways to face circumstances courageously that the older thick conception never envisioned" (Lear 2006, 65). By analogy, I am suggesting that because of the unfolding climate catastrophe at least some of our conceptions of human environmental excellence may likewise have to be thinned out before re-emerging in new and culturally novel forms.

Lear reconstructs Plenty Coups's deliberation as the chief:

> We certainly know that we cannot face the future in the same way that we have been doing.... We must do what we can to open our imaginations up to a radically different set of future possibilities.... If I am to go on living, I need to be able to see a genuine, positive, and honorable way of going forward. So on the one hand, I need to recognize the discontinuity that is upon me—like it or not there will be a radical shift in form of life. On the other, I need to preserve some integrity across that discontinuity.... My commitment to goodness is manifested in my commitment to the idea that something good will emerge even if it outstrips my present limited capacity for understanding what that good is (Lear 2006, 93).

Plenty Coups was committed to a goodness that transcended his understanding and he was hopeful his people would be able to "get the good back"—not only would they survive the destruction of their traditional forms of life but they would return again to flourish in the presently unimaginable new world.

Plenty Coups embodied a virtue Lear calls radical hope, which is "basically the hope for revival: for coming back to life in a form that is not yet intelligible," the virtue of seeing that the world's goodness outstrips the ability of one's culture to capture it (Lear 2006, 95). Radical hope is against despair, even in the face of a well-justified despair. It is the idea that an inadequate grasp of the good should not lead one to believing it is not to be hoped for.[5] "At a time of radical historical change, the concept of courage will itself require new forms. This is the reality that needs to be faced—the *call for concepts*—and it would seem that if one were to face up to such a challenge well it would have to be done imaginatively," with what Lear calls "imaginative excellence" (Lear 2006, 118).

[5] I have been influenced in my description of radical hope by listening to Jonathan Lear and Hurbert Dryfus discuss the work in an author-meets-critics session at the 2007 Pacific Division meeting of the American Philosophical Association.

Today we need imaginative excellence in the conception of new cultural patterns, new human forms of life that will allow us to live well on a planet that may be quite unlike the one under which human civilization has developed and all life on Earth is adapted. How might we answer this call for concepts? Radical hope is one response, a product of imaginative excellence that allows courage to be manifest in situations where one has an outdated conception of living well. Courage traditionally concerns the willingness to risk significant harm defending a worthy good, but radical hope is a form of courage at the end of goodness, underpinning action on the mere hope that someday the good will return in a presently unimaginable form. Environmentalism tends to attribute some form of goodness to the natural world. Today, as we bring about the "end of nature," are we committed to the possibility of a new and historically novel form of goodness in nature?[6] For the sake of the possibility that they, too, may lead flourishing lives, I believe we owe such courage to future generations.

Virtues of the Future

How does a conception of goodness in the natural environment connect with human flourishing? On my view, a virtue is a character trait that a person needs to live well in the sense that possession of the virtues is constitutive of human flourishing. The particular version of neo-Aristotelian eudaimonism I endorse (but won't defend here) adopts the *natural goodness approach*, according to which evaluative judgments of human character and rationality depend "directly on the relation of an individual to the 'life form' of its species" (Foot 2001, 26).[7] Environmental virtues, then, are excellences regarding various relations to the organisms and ecosystems of the terrestrial biosphere where our specifically human lives unfold. Environmentally good human beings recognize what is fine in and about the natural world and are habitually disposed to act well regarding these values, thus enabling the goodness of nature to have a substantive role in human flourishing, just as the goodness of friends and human community—via the relevant interpersonal virtues—can have such a role.

This bare sketch of a theory of environmental virtue provides a framework for considering how the specific content of some environmental virtues may reflect significant cultural and environmental change. Consider also a preliminary distinction between character traits suited for living well *through* a time of radical cultural and environmental change (what I'll call "virtues of transition") and what human excellence may require once these crises have subsided (or, "virtues of the future").[8] In this section I'll outline a place for radical hope in some environmental virtues of transition and identify two parts of a theory of environmental virtues of the future.

[6] "The end of nature," is an allusion to (McKibben 1989), which I discuss in the next section.

[7] I outline and defend a version of the natural goodness approach in Thompson (2007).

[8] By analogy, we may distinguish between forms of excellence in the midst of an unfolding but more ordinary catastrophe, like the sinking of the Titanic, and what excellence requires in the aftermath.

If radical hope is for revival of the good, when the goodness one understands is no longer possible, we may ask what culturally-bound conceptions of *environmental* goods are threatened by impending changes. I'll discuss environmental goods along two dimensions, natural value and lifestyle value. The issue of losing a conception of specifically natural value is closely associated with McKibben's view; if we value the natural in so far as it is untouched by humans but anthropogenic climate change brings an end to this world, then this form of goodness in nature will be lost. A second culturally-bound conception of goodness connected with the natural environment is more mundane, but perhaps more common: modern expectations of comfort and convenience, the high "standards of living" that characterize the felt preferences of most members of our consumer society, are made possible by an abundance of natural resources, sinks, and ecosystem functions. A relatively stable and biologically diverse natural environment continues to make possible the good we find in our consumer and materialist lifestyle. Both conceptions of environmental goods are endangered by the emerging climate crisis. Thus, a new form of an inherent goodness in nature and a new conception of living well are relevant objects of radical hope for the environmentalist.[9]

Radical hope, as courage manifest in environmental virtues of transition, is primarily against despair and hopelessness, which the apparently intractable problem of climate change can supply in spades. Thus radical hope can be realized in commitment to the political and social struggles connected with environmental activism. What must be resisted, without being denied, is that very significant harms, including degraded ecosystems, mass species extinctions, and considerable amounts of human injustices and suffering, are now inevitable. It seems likely that even our best efforts will not entirely prevent such evils nor are our efforts likely to make a significant difference for many years. Yet we must not give up resistance and the struggle for change. Character dispositions against a reluctant acquiescence stem from radical hope, as a form of commitment against what may be justifiable despair. We can all find evidence in our associations with others to support this feeling of inevitability—it seems that enough people simply will not voluntarily make the kind of changes in lifestyle or social organization required to effect significant mitigation. Neither does it seem that political responses will be sufficiently prompt or aggressive. Still, we must be actively committed, with a radical hope, to the belief that cultivating virtues of environmental activism will not be in vain, ironically in spite of all strong evidence to the contrary.[10]

Additionally, mitigating the damages of climate change will require personal commitment to the struggle against pervasive materialist dispositions, which effectively characterize all of us as members of an industrialized consumer society. Here again radical hope is manifest as a kind of commitment, in restraint and a

[9] It is likely that "consumerism," as commonly understood, involves a faulty conception of the good. A conception of the good as a life of material wealth has long been the target of philosophical criticism. In the context of this paper "consumerism" is not meant to pick out only some subset of our society but to characterize the whole of it. What is at risk is what many of us conceive to be, mistakenly or not, at least part of the good life.

[10] On the role forgiveness can play in overcoming psychological barriers to environmental activism (see Norlock 2009). On the virtues of environmental activism, see (Sandler 2007, 49 ff).

resolve against taking too much satisfaction in the consumptive modes to which we are accustomed. As with any addiction, we must have a radical hope that what begins as mere continence and stoicism will eventually flower into true virtue, as the constellations of our pleasures and pains are realigned. While it is one thing to intellectually understand strong evidence supporting the view that materialistic dispositions are inversely correlative with subjective well-being, it is still quite something else to recognize many of one's own seemingly ordinary activities or expectations, e.g., a hot shower every day or fresh fruit out of season, may involve an excessive level of indulgence, when set against the requirements of ecological sustainability.[11]

Also, it's easy to identify *other* people, for example those who often buy things on impulse, as overly consumptive; of course, statistical outliers exceed the consumptive norm. Yet, it is less common to recognize that the statistical *norm* of consumer society already involves excesses, and far more psychologically difficult to recognize that you yourself live closer to the norm than justice may permit. For many of us, to even conceive the possibility of living well with significantly less requires a kind of imaginative excellence similar to that of Plenty Coups. Then, actually *living committed* to this vision will require the resources of a radical hope, helping one to swim against the currents of dominant culture. Radical hope is manifest in the virtues of commitment to a new form of living well, in the *resolve* to live well with less, along with the *ingenuity*, *frugality*, and *restraint* required to accomplish this.

For the sake of improving the prospects of future generations to flourish, we must have radial hope that good human lives do not require the material conditions and amount of energy required to support contemporary standards of living. While it's not difficult to *believe* that this is true—a strong vein of romanticism for the simple life lurks nearby—it's very difficult to actually *live* like it's true. Analogous to Mill's insight that individual liberty enables various "experiments in living" from which society as a whole benefits, the world's well-to-do need to begin new experiments in living, working out how to recognize and live a good life while consuming substantially less, thus undertaking experiments in simplicity. A real commitment to this takes a radical hope because in the dominant world culture we *just don't have* an alternative yet viable conception of the good life. But success here will both help mitigate environmental damage and position us to share what we learn with our successors, who will surely *have* to live with less.

At the same time, we cannot stop developing and consuming the world's natural resources; we must live *from* the world (O'Neill et al. 2008, 1). Undeniably, the non-human world of nature has instrumental value for human life and well being. Yet many environmentalists are committed to the idea that some parts of nature are also valuable non-instrumentally, bearing some form of intrinsic goodness. Whether evaluative properties exist independently of evaluators or not, there is a long tradition of finding intrinsic value in nature as unsullied by human intervention. Without denying that the term "natural" is notoriously ambiguous, Dale Jamieson

[11] On the evidence on consumptive dispositions and their relation with subjective well being, see (Sandler 2007, 55 ff).

tells us "the sense of naturalness that is important for many who value nature is this: something is natural to the extent that it is not a product of human influence" (Jamieson 2008, 162).

McKibben has famously argued that anthropogenic climate change is effectively the end of nature, on the grounds that climate change affects every part of the Earth's biosphere and thus human interference with the climate system leaves no part of the biosphere free of human influence (McKibben 1989). We can allow that "natural" is a matter of degree, acknowledging there is a natural world beyond social construction but still recognizing that much of what we tend to think of as natural is, in fact, significantly influenced by human activity.[12] Jamieson rejects McKibben's claim that global warming is the end of nature on the basis that naturalness comes in degrees, but he reinterprets McKibben's anxiety as a moral worry about the loss of nature's autonomy, what he characterizes as nature's effective freedom from human domination.

According to Jamieson, while not the *end* of nature as independent from humans, climate change poses a serious threat to the autonomy of nature (Jamieson 2008, 166; see also Katz 1997; Heyd 2005). How is the independence of nature distinct from its autonomy? The idea of nature as independent from human beings is an ontological one, a metaphysical thesis that depends on a false dualism where any human influence at all transforms the very being of something from nature into artifact. On the other hand, the idea of nature as autonomous is fundamentally a thesis that human beings are neither causally nor morally responsible for the basic conditions or operations of the Earth's biosphere. So distinguished, we can recognize that massive anthropogenic greenhouse gas emissions will not alter the basic metaphysics of nature (whatever they are) but such intervention can and does count against the causal autonomy of nature and in favor of a collective human causal and moral culpability.

As a result of many human activities including those that result in atmospheric forcings, our managing a significant reduction thereof, or the possible effects of intentional climate change, i.e., geo-engineering, we can no longer believe in good faith that the Earth's climate and large-scale biological systems carry on with autonomy in the relevant sense. The Earth's biosphere will now forever be indelibly marked, or determined, by human activity. We are now in what some are beginning to call the Anthropocene epoch (Crutzen 2002). If this is right, then radical hope today must take the form of commitment to the possibility of a new conception of non-instrumental, and not simply aesthetic, natural goodness—a form of value in nature essentially *distinct* from the value of nature's autonomy.

If we abandon commitment to an intrinsic value of naturalness in the autonomy of nature, intrinsic value theorists could still defend teleological centers of life pursuing their own good in their own way or individual sentient beings as the loci of human-independent value in nature (Taylor 1986; Singer 1975; Regan 1983). Indeed, some or all living things may have real evaluative properties; they may be valued or actually valuable for their own sake, non-instrumentally. Of course, simply as *moral agents* we should not ignore morally relevant, non-human values

[12] For example, see the discussion of the red-cockaded woodpecker in (Thompson 2009).

and, further, there remain difficulties answering how these values connect with moral reasons (Nolt 2006). But it's not at all obvious how these non-human sources of value bear on us as specifically as *environmentalists*. Could valuing naturalness, i.e., the autonomy of nature, be essential to modern environmentalism?

Consider for example the non-human value of artificial forms of life, perhaps some of them conscious, created someday in a laboratory on a spaceship. We probably do bear moral responsibility regarding them, perhaps owing them duties directly on account of their moral considerability, but these moral relations do not fit easily into any conception we have today of an *environmental* ethic, if part of the basic axiology of environmentalism is reverence for nature's autonomy. As a source of value, the intrinsic values associated with life and sentience seem to face the same problem as the value of biodiversity: promoting it can conflict with protecting naturalness and in the end "[w]hat most environmentalists want is naturally produced biodiversity," and will favor naturally produced forms of consciousness and life over artificial alternatives (Jamieson 2008, 168). Naturalness in the relevant sense *is* essential to environmentalism today. On Jamieson's view, "without placing some value on the autonomy of nature, there is really nothing left to environmentalism as it has come to be constructed."[13] If this is right, then the climate crisis can also be seen as a crisis for contemporary environmentalism. My analysis supports the idea that environmentalism in the future, including environmental virtue theory, will hold a significantly diminished place for valuing the good of autonomy in nature.

On the account I have provided, to deny the autonomy of nature is to affirm some conception of human responsibility. Widespread scientific consensus affirms human *causal* responsibility for global climate change[14] yet our received paradigms of moral responsibility are ill suited for responding to the structure of collective action that is characteristic of many environmental problems, including anthropogenic global warming. By hypothesis, we cannot provide culturally thick accounts of some environmental virtues of the future because today we lack the requisite concepts. But considerations against valuing nature as autonomous support the idea that moral responsibility will play a central role in tomorrow's environmentalism. I offer two related suggestions. The first concerns a specific existential burden and the second concerns understanding *being responsible* as a virtue, including the place that related character traits could have in practices of adaptive ecosystem management and ecological restoration.

When nature is *not* autonomous from humanity, human beings are morally responsible in an unprecedented way—collectively responsible for the parameters of global climactic conditions and, thereby, for large-scale biological and meteorological phenomena (e.g., extinctions, hurricanes). As such, each member

[13] In personal correspondence on 28 January 2007. As another example, Philip Cafaro writes: "The [Crow] analogy with environmentalists might go like this: 'Maybe we will be able to save our own hides and somehow figure out ways to survive, maybe even with some of our dignity and ability to do good intact. But the things that we really want to save as environmentalists, like wild nature? Forget it! Like the buffalo, they're on the way out.' But if this is all we can hope for ("radical hope"), then I for one might rather just throw in the towel" (personal correspondence on 17 November 2008).

[14] See the Intergovernmental Panel on Climate Change, AR4.

of humanity bears some of this responsibility or is at least potentially blameworthy (French 1998). Yet, when nature *was* autonomous, human beings were free from the existential burden of bearing moral responsibility for the some of the basic conditions supporting all life on Earth.[15] Bearing this burden may be understood as a kind of natural bad, whereas the autonomy of nature and subsequent human moral freedom was a natural good. If such events cannot be avoided, then it is better if we are not responsible for the meteorological or other climatic phenomena that—combined with social conditions—cause significant harm to humans and other life. If responsibility for global climate involves the loss of this way of finding instrumental value in nature, we may form the radical hope that our responsibility for nature, rather than being an unwelcome burden, could somehow fit into a conception of natural goodness that we don't yet understand. We are, after all, a part of nature.

The second suggestion begins by bringing together two ideas central to one of Jamieson's early articles on the ethics of climate change: the need to develop new conceptions of responsibility and the need to focus more on character (Jamieson 1992). Jamieson's worry is that our received conception of responsibility "presupposes that harms and their causes are individual, that they can readily be identified, and that they are local in space and time" (Jamieson 1992, 148). However, features of global environmental problems like climate change diverge from this paradigm. Recent work on collective and shared responsibility may begin to attenuate this line of thinking but indeed what is generally absent from the literature is discussion of responsibility as a virtue.[16]

Standard analyses of moral responsibility focus on metaphysical questions about the nature of moral agency or with the ascription of moral praise or blame and consequently the administration of just deserts. Yet we also praise both individuals and collectivities for *being responsible*, remarking not on the satisfaction of some particular duty but on a worthy and good condition of character involving features such as reliability, judgment, and initiative in the context of mutual accountability under a plurality of normative demands (Willimas 2008, 457; see also McKeon 1957). While there remain difficult questions about the distribution of collective responsibility to individuals, the suggestion here is that a significant part of human environmental excellence in the Anthropocene will involve character traits disposing individuals to act well at all times as agents both causally and morally responsible for the conditions of all life on Earth. I predict that the virtues of being environmentally responsible will extend beyond dispositions toward conservation to include some conception of how an active human element is itself a constitutive part of the goodness in a non-autonomous natural world.

Related ideas are beginning to appear in the literature on adaptive ecosystem management, specifically concerning the practices of ecological restoration, where

[15] Martha Nussbaum remarked that one central disanalogy between our situation today and the plight of the Crow is that the Crow were in no way responsible for the utter catastrophe that befell them.

[16] One notable exception is (Willimas 2008). Williams notes that "the question of whether adults of sound mind are responsible by virtue of, say, free will does not help with the question of how some better exemplify responsibility than others, nor how collective bodies might manifest [the virtue of] responsibility" (Willimas 2008, 458 footnote #6).

debates concerning the value of human intervention with nature are commonplace. As part of a broader social practice, Andrew Light argues, participation in ecological restoration can have a significant positive impact on the *relationship* between human culture and nature, a relation he defends as having normative content. The practice of ecological restoration has a central role in what Light calls the restoration of a culture of nature (Light 2000). The form of environmental goodness, which he identifies here as "the task of a more pragmatic environmental philosophy," includes recognizing a non-instrumental value in relation to nature that is distinct from nature's autonomy, thus fitting the bill (on my account) as an appropriate object for radical hope, in light of which specific environmental virtues of the future, including being responsible, may be measured.

Also along these lines, William Throop advocates a virtue approach to restoration, arguing that the central metaphor should be healing, rather than gardening or engineering. The virtues of the healer, including humility, restraint, sensitivity, and respect, will lead practitioners to avoid hubris when actively engaging in the co-determination of natural systems ("Environmental Virtues and the Aims of Restoration," unpublished manuscript). Finally, Ronald Sandler argues that the realities of global warming will modify the substantive content and relative importance of the virtues associated with restoration and assisted recovery, including those related to "openness and accommodation" and raising the salience of virtues connected with "reconciliation" ("Restoration, Virtue, and Global Warming," unpublished manuscript).

To review, I have argued that if we face significant cultural change, then some culturally thick conceptions of the environmental virtues—including an understanding of what it is to be environmentally responsible—will be up for revision *and* that we will soon face significant cultural change. I have identified how anthropogenic climate change threatens at least two different culturally-bound conceptions of environmental goodness, how radical hope should be part of the commitment to virtues of environmental activism, how it must be mixed with imaginative excellence in the virtues of living lightly and re-visioning what counts as a high standard of living, and how environmentalists should have radical hope in a form of non-instrumental natural goodness, a value beyond the autonomy of nature, which recognizes human causal and moral responsibility. Lastly, I identified where the virtues of being environmentally responsible are beginning to emerge in one strain of pragmatic environmental philosophy.

Technology and the Ghost Dance

In closing, I'll briefly consider the objection that a source of alternative clean energy will enable us to avoid, for the time being, significant changes in the dominant consumer culture. This possibility entails that a civilization built on exploiting a non-carbon-based form of energy is consistent with the on-going existence of our culture of material consumption.

Daniel Nocera at the Massachusetts Institute of Technology envisions a world of decentralized solar energy production as an achievable energy future (Revkin 2008).

According to Nocera, the energy draw of all current human activity is equivalent to approximately 13 terawatts watts (=13 *trillion* watts) and by 2050 it will be thirty terawatts. Meeting our current energy needs with fossil fuels is obviously unsustainable, so how could we supply the additional 17 terrawatts without coal, oil, or gas? Using all plant life on Earth would produce seven terrawatts, building a new nuclear plant every other day until 2050 would yield eight, harvesting all wind energy available on Earth's surface gives two, and "you get one more terawatt if you dam every other river on the planet" (Revkin 2008). This hybrid package of hydrocarbons and "clean" alternatives would satisfy the expected demand but avoids no environmental problems, while adding many others.

However, the sun delivers eight hundred terawatts to Earth continuously and, without a substantial change in lifestyle, by 2050 we will need only thirty of those. But we do not currently have the technology to *capture* what we need. We would have to develop and deploy such technology globally within a very short time to avoid catastrophic and runaway climate change, yet we devote thirty times the financial resources to health research than we do to basic energy research and development (Revkin 2008). Under these conditions, what reason is there to hope for a silver bullet of clean alternative energy?

Still, many people have this hope and I won't say they can't be right. In any case, if the dream of alternative energy prevents us from considering the currently less palatable, but perfectly possible futures, then we should be sorry. An important part of courage, on Lear's analysis, is facing the realities one would otherwise want to deny. For example, Sitting Bull, chief of the Sioux, embraced another dream vision of a messiah who would strike down the white settlers and usher in a return of the old ways. According to Lear, Sitting Bull's embrace of the Ghost Dance was a failure to engage reality (Lear 2006, 150).[17]

Wishing only for an alternative energy solution, instead of making preparations for cultural change, in part by revising conceptions of the value of nature under anthropogenic climate conditions, is analogous to Sitting Bull's embrace of the Ghost Dance. Both are cases of what Freud called "turning away from reality" (Lear 2006, 151).[18] Just as rituals like the Ghost Dance constituted a traditional response for the Sioux, so we are accustomed to expecting that technology will provide a miraculous solution. Since necessity—or perhaps in this case, scarcity—is the mother of invention, in the face of global environmental catastrophe human ingenuity will surely hit upon a technological solution to save us all the trouble of significant behavioral and conceptual change. Or so goes one telling of how things always seemed to work in the past.

[17] "It is the hall mark of the wishful that the world will be magically transformed—into conformity with how one would like it to be—without having to take any realistic practical steps to bring it about" (Lear 2006, 150).

[18] In such cases, "[s]ymbolic rituals, [perhaps like shopping]... take over life—whether in an individual's private life or in the group activity of a culture—and they become a way of avoiding the real-life demands that confront one" (Lear 2006, 151).

Acknowledgments Earlier versions of this paper were presented at the *Human Flourishing and Restoration in the Age of Global Warming* conference at Clemson University, the *International Society for Environmental Ethics* during the 2009 *American Philosophical Association* (Central Division), and the Inland Northwest Philosophy Conference on *The Environment* at the University of Idaho. I am grateful to those audiences and to Jeremy Bendik-Keymer, Philip Cafaro, Baylor Johnson, Jason Kawall, Andrew Light, Kathryn Norlock, Martha Nussbaum, Ronald Sandler, Sarah Wright, and three anonymous referees for this journal for helpful comments.

References

Cohen, J. E. (2005). Human population grows up. *Scientific American*. Accessed February 12, 2009, from http://www.sciam.com/article.cfm?id=human-population-grows-up.

Crutzen, P. J. (2002). Geology of mankind. *Nature, 415*(23). Accessed March 9, 2009, from http://www.nature.com/nature/journal/v415/n6867/full/415023a.html.

Dillard, Annie. (1998). The wreck of time: Taking our century's measure. *Harper's Magazine, 296*, 51–56.

Foot, P. (2001). *Natural goodness*. New York: Oxford University Press.

French, P. (Ed.). (1998). *Individual and collective responsibility*. Rochester, VT: Schenkman.

Gartner, John. (2008). "Glaicers: Earth's early warning system". Accessed August 10, 2008, from http://www.chasingglaciers.com/features.cfm.

Goodwin, N. R., Nelson, J. A., Ackerman, F. & Weisskopf, T. (2007). Consumer society. *Encyclopedia of Earth*. Cutler J. Cleveland (Ed.). Washington, DC: Environmental Information Coalition, National Council for Science and the Environment. Accessed August 10, 2008, from http://www.eoearth.org/article/Consumer_society.

Hansen, J., Sato, M., Kharecha, P., Beerling, D., Masson-Delmotte, V., Pagani, M., Raymo, M., Royer, D. L., & Zachos, J. C. (2008). Target atmospheric CO_2: Where should humanity aim? Accessed March 30, 2009, from http://www.columbia.edu/~jeh1/2008/TargetCO2_20080407.pdf.

Heyd, T. (Ed.). (2005). *Recognizing the autonomy of nature*. New York: Columbia University Press.

Jamieson, D. (1992). Ethics, policy, and global warming. *Science, Technology & Human Values, 17*(2), 139–153.

Jamieson, D. (2008). *Ethics and the environment*. New York: Cambridge University Press.

Katz, E. (1997). *Nature as subject: Human obligation and natural community*. Lenham, MD: Rowman & Littlefield.

Lear, J. (2006). *Radical hope: Ethics in the face of cultural devastation*. Cambridge: Harvard University Press.

Light, A. (2000). Ecological restoration and the culture of nature: A pragmatic perspective. In P. H. Gobster & B. Hull (Eds.), *Restoring nature*. Washington, DC: Island Press.

McKeon, R. (1957). The development and the significance of the concept of responsibility. *Revue internationale de philosophie, 11*(39), 3–32.

McKibben, Bill. (1989). *The end of nature*. New York: Random House.

Monbiot, George. (2008). One shot left. Accessed March 15, 2009, from http://www.monbiot.com/archives/2008/11/25/one-shot-left/.

Nolt, John. (2006). The move from good to ought in environmental ethics. *Environmental Ethics, 28*(4), 355–374.

Norlock, Kathryn. (2009) Forgiveness, pessimism, and environmental citizenship. *Journal of Agricultural and Environmental Ethics 23*(1).

O'Neill, J., Holland, A., & Light, A. (2008). *Environmental values*. New York: Routledge.

Regan, T. (1983). *The case for animal rights*. Berkeley: University of California Press.

Revkin, A. (2008). All energy roads lead to the sun. *New York Times*, 27 March 2008. Accessed August 10, 2008, from http://dotearth.blogs.nytimes.com/2008/03/27/all-energy-roads-lead-to-the-sun/.

Sandler, R. (2007). *Character and environment*. New York: Columbia University Press.

Singer, P. (1975). *Animal liberation: A new ethics for our treatment of animals*. New York: Random House.

Taylor, P. (1986). *Respect for nature: A theory of environmental ethics*. Princeton, NJ: Princeton University Press.

Thompson, A. (2007). Reconciling themes in neo-aristotelian meta-ethics. *Journal of Value Inquiry, 41*(2), 245–264.
Thompson, A. (2009). Responsibility for the end of nature, or: How I learned to stop worrying and love global warming. *Ethics & the Environment, 14*(1), 79–99.
Willimas, G. (2008). Responsibility as a virtue. *Ethical Theory and Moral Practice, 11*(4), 455–470.

J Agric Environ Ethics (2010) 23:61–83
DOI 10.1007/s10806-009-9190-5

ARTICLES

Species Extinction and the Vice of Thoughtlessness: The Importance of Spiritual Exercises for Learning Virtue

Jeremy Bendik-Keymer

Accepted: 11 May 2009 / Published online: 14 June 2009

Abstract In this paper, I present a sample spiritual exercise—a contemporary form of the written practice that ancient philosophers used to shape their characters. The exercise, which develops the ancient practice of the examination of conscience, is on the sixth mass extinction and seeks to understand why the extinction appears as a moral wrong. It concludes by finding a vice in the moral character of the author and the author's society. From a methodological standpoint, the purpose of spiritual exercises is to create a habit of thoughtfulness in the *writer*, and by way of teaching, to suggest one to the reader. Such a habit is important, at least, because virtue is a habit. In other words, there can be no learning of virtue itself without habituation into it. Accordingly, I frame the sample spiritual exercise with a deliberately controversial objection to contemporary academic virtue ethics and with a justification for why the spiritual exercise is important for taking virtue ethically. And I end the paper with some further remarks explaining the form of the exercise and its relevance to doing philosophy. In this way, the paper makes and illustrates a methodological point about virtue ethics based on a meta-ethical assumption about virtue as a habit, and it does this by focusing on a pressing environmental problem in the twenty-first century.

Keywords Species extinction · Sixth mass extinction · Virtue theory · Spiritual exercises · Meta-ethics · Meta-philosophy

> The purpose of our examination is not to know what virtue is, but to become good.
> —Aristotle (1999), *Nicomachean ethics*, 1103b28-9

J. Bendik-Keymer (✉)
Le Moyne College, 616 DeMong Drive, Syracuse, NY 13214, USA
e-mail: keymerjd@lemoyne.edu

Discourse is a privileged means by which the philosopher can act upon himself and others. … It is always intended to produce an effect, to create a *habitus* within the soul, or to provoke a transformation of the self.
—Pierre Hadot (2002), "Philosophy and philosophical discourse", p. 176

Humanity has initiated the sixth great extinction spasm, rushing to eternity a large fraction of our fellow species in a single generation.
—E. O. Wilson (1999), *The diversity of life*, p. 32

Learning Virtue Without Learning It

I worry that, paradoxical as it is, we[1] teach virtue without teaching it. If virtue is above all a habit, then to teach it without habituation is not to teach it. This is my first worry over academic virtue ethics—a disparity between form and content.[2]

In this essay, I will frame a way of learning virtue that attempts to meld form and content—to have each support the other. The way of learning draws on an ancient philosophical tradition that has been translated as the tradition of "spiritual exercises"—from the Greek *askesis* (the root of our word "ascetic"). However, we should not think of this tradition as the tradition of ascetic practices, but as the tradition of methods by which philosophy becomes an *ethos*—a way of life, our character. I will provide an example of this ancient way of learning, and my frame will allow us to be aware of the theoretical assumptions behind its methodology.[3]

Are we concerned with the *concept* of virtue, or with virtue itself? Meta-ethics, whose point is to understand the nature of ethics, is understandably after the former. As such, its efforts fall squarely within the domain of theoretical reason, whose point is truths and whose value is the true. However, ethics—the part of everyday life, not the field of study—are squarely practical. They fall within the domain of practical reason, whose point is doing something in the world and whose value is the good (or, for some, the right). If we are concerned only with meta-ethics, we can rest content with the concept of virtue. But, even so, we cannot thereby assume to be learning ethics. Rather, to learn ethics, we need to *take* the concept of virtue *ethically*, which is to subsume our theoretical pursuits under practical reason, the

[1] Mainstream, academic virtue ethicists, and those interested in the *point* of mainstream, academic virtue ethics. I believe the mainstream approach to virtue ethics could be related to one reading of Aristotle's *Nicomachean Ethics*, although its causes might be more contemporary in the dynamics of university research, intellectual status, and markets. That reading of Aristotle pictures him as the first major virtue *theorist*. But that reading of Aristotle is mistaken. In what follows, I will rely on an *alternative* reading of Aristotle, according to which his reflections are not divorced from habituation, but part of a citizenry's attempt to come to grips with virtue. The goal on this alternative reading, as Aristotle himself said, is to *become* virtuous, not simply to know virtue. See Aristotle (1999), pp. 1103b28–1103b29. See also the powerful passage at 1105b14–1105b19.

[2] It is the *first* problem, because it displays a grasp of the *point* of virtue, and so of teaching virtue. Virtue is to be practiced, above all. That is its point. Without aiming to habituate people into virtue, academic virtue ethics is not lined up right *from the beginning*.

[3] I came to my approach through reading Pierre Hadot (2002) and my example of a spiritual exercise is particularly indebted to the ancient idea of the examination of conscience which Hadot discusses (pp. 198–202).

primary form of reason in matters ethical. That is to become concerned with virtue itself.[4]

My point is that someone concerned ethically with virtue will be concerned with virtue itself. The point of virtue is to live excellently, or well, and, for a person concerned with being ethical, the point of investigating virtue is not to know virtue, but to *become* virtuous.[5] That Aristotle took this as self-evident shows he took his investigation ethically. The question then is whether we can do philosophy of virtue without taking virtue ethically. *That* is a question about the nature of philosophy, the love of wisdom. I will consider it only briefly here. For my purposes, I will discuss virtue from the standpoint of taking it ethically.

Now for someone concerned with virtue itself, the forms of life that build virtuous habits are as crucial as any point of theory.[6] This conclusion implies that philosophers concerned with virtue itself should shape the styles of their theoretical discourses so that they contribute to the formation of virtuous habits.[7] Doing so is a matter of philosophical integrity, not simply a welcome ornament to philosophical work on virtue itself. I attempt to address this stylistic challenge to our integrity as virtue ethicists through my essay in environmental virtue ethics. I express not only a specific, virtuous practice, but a methodological point about how we should be doing virtue ethics if we are concerned with virtue itself. That point flows from a meta-ethical assumption about the kind of thing virtue is: a state of character expressed by habits of mind and action.

In what follows, I will present a sample, virtuous practice in which form and content cohere. My way of doing so is to frame it for academic virtue ethicists so that it is clear why the style of the practice is what it is. By way of framing, I first discuss philosophy's relation to theory and then how virtue itself exceeds theory. This second section addresses my assumption about virtue being a habit. There then follows the sample virtuous practice, what some historians of ancient philosophy call a "spiritual exercise" (Hadot 2002, pp. 189–190).[8] The example finished, I conclude with a section on the audience and style of the exercise, underlining how its form and content cohere. In particular, I focus on the idea of the form of

---

[4] I thank Chin-Tai Kim and Colin McLarty for discussion leading to conclusions of this paragraph.

[5] Again, see Aristotle (1999), pp. 1103b28–1103b29: "The purpose of our examination is not to know what virtue is, but to become good." Being good, for Aristotle, is being virtuous, as his function argument from book 1 of the *Ethics* makes clear, cf., 1098a17.

[6] They might be *more* important if, as Aristotle sometimes suggests, we cannot think straight without virtuous habits. The point is not that hard to imagine: an intemperate person will *not* think straight around bodily pleasures. See Aristotle (1999), pp. 1144a29–1144b1 and pp. 144b31–144b32.

[7] See my review of Hadot's *What Is Ancient Philosophy?* (Bendik-Keymer 2008). Although I do not have the space to go into the argument here, I would like to say that philosophers—not just those interested in virtue—should check to see that their styles are appropriately practical, for at least a good part of philosophy's object is a wise life, just as virtue's object is a life lived well. That is, practical reason appears to be central to philosophy, even though we may associate philosophy with theory. Currently, I am exploring this idea in other work through philosophy's relation to what I call the "grammar of being." The grammar of our being shows us that theoretical, practical and what I call "relational" reason are co-primary. Being is meaningless without the logic of any one of these kinds of reason, and philosophy is pointless then, too, since I conceive of philosophy as pursuing our full being.

[8] The technical term is *askesis*—the Greek root of our "ascetics." See also Hadot (2003) and his magisterial study of Marcus Aurelius's *Meditations* and its relation to Epictetus's teaching (Hadot 1998).

questioning, which allows us to distinguish the point of reflection from effects of reflection and a manner of writing from a manner of reading, and which underlines how learning virtue itself is primarily the work of practical, not theoretical, reason.

Philosophy's Relation to Theory

I do not have space to discuss fully whether we can do philosophy of virtue without being concerned with virtue itself, but I suspect that we cannot.[9] Even so, it will help to discuss philosophy's relation to theory in a preliminary way.

Philosophy is the love of wisdom, and—in its ancient form—its point was to become wise.[10] Although I think there are reasons to criticize this ancient formulation, it does pose a helpful challenge that forces us to be clear: the question is whether theory, and in general *the theoretical life*, seeks the same goal as philosophy and *the philosophical life*? If they do, then the life of a theoretician seeks to become wise above all, and the practice of theory has in view wisdom in all instances. The conclusion is a simple consequence of the relation of means to their ends, with the end being wisdom. Does the life of theory seek wisdom above all else?[11]

The answer to this question requires a definition of theory and a definition of wisdom. On my understanding, theory is comprehensive and abstract knowledge of things. By contrast, wisdom reflects our full being in all its dimensions.[12] Our full being involves truth, goodness, and beauty—the values for knowledge, action, and connection. Theory is at most a third of wisdom, then. It aims at knowledge and its

[9] To see why the matter is difficult, think about how it could be wise to *not* take virtue ethically. Suppose, for instance, one's system of virtues were predicated off of an unjust ethical system. Then it would be wise to *refuse* to take virtue ethically. It would be wise to stand back and to look at virtue theoretically and to refrain from practicing it until one has extirpated the injustice from one's system of values. That being said, it seems unlikely to me that we could approach virtue philosophically without trying to learn virtue itself. At the very least, this is because not much of life is exceptional in the sense just discussed. Most life demands what we take to be virtue unproblematically. Moreover, in the previous example of an unjust ethical system, being *truly* just seems to be necessary for adequately criticizing that system. In other words, the theoretical suspension of virtue itself depends on virtue itself.

[10] See Hadot (2002), pp. 220–231 and pp. 42–50, where, for Socrates, wisdom appears is an unattainable *regulative ideal* driving on onward (the Kantian language is my own).

[11] My main reason for criticizing the ancient conception of the point of philosophy is that I think it privileges practical reason. As I alluded to briefly before, I think theoretical, practical, and relational reason are co-primary in the grammar of our being, and it does not make sense to me that the pursuit of wisdom would distort the grammar of our being. But if it won't, then the goal of philosophy should not be primarily practical. Rather, it should be pluralistic with respect to kinds of reason. In other words, the goals (plural) of philosophy should be implicit in wisdom (singular), the many in that one, and these goals should reflect the grammar of being—a wise life (the practical reason of philosophy), truth (the theoretical reason of philosophy), and harmonious connection with others (the relational reason of philosophy). I hope to explain these points more fully in future work.

[12] The expression "full being" comes from Irad Kimhi.

value is truth. But wisdom exceeds theory. It is practical, too. And it involves connecting with others in a way that is not reducible to knowing about them or knowing how to act alongside them. Theory is a *part* of wisdom.[13]

Once you realize this logical point, it is obvious that a theoretical life needn't be a wisdom-seeking one unless it is taken philosophically. Whereas wisdom implies an acknowledgement of theory's place in our full being, one could pursue theory without pursuing wisdom, a point that is all too obvious in the twentieth century caricature of the scientist who pursues truth but forgets his humanity or in the caricature of the analytic philosopher who pursues truth but is completely out of touch with both social justice and others. So, if you are a theorist trying to do philosophy, the question to ask yourself is whether you aim to become wise. If you do, then you do philosophy, too. If not, then your activity isn't philosophical. A material biconditional obtains between aiming to become wise and philosophy, but theory is at best a necessary and not a sufficient condition of wisdom.

Virtue Taken Ethically

In the context of being ethical, my worry about theory is not about its uselessness, but about its *pointlessness*. Theory can certainly be useful. But theory seeks the true. Ethics, by contrast, seeks the good (or, for some, the right). The former concludes in truths (or realizations that truths are not forthcoming, a move in the space of the true). The latter concludes in deeds. If I am trying to become ethical, to approach virtue merely theoretically is to make a category error, and—from an ethical standpoint—a grave mistake. Another way to put this concern is, if I want to learn the wisdom of doing the right thing, it won't be enough for me to merely think about my character.

How do we take virtue ethically? Think about what virtue itself is. If virtue itself is X, then a discourse that does not promote X does not promote virtue itself. Let us say, then, that virtue itself is a habit. Then a discourse that does not promote virtue as a habit does not promote virtue itself. You'd have to promote virtue's habits to take virtue ethically.

Should we, then, regard virtue as a habit? For the most part, I follow Aristotle here. Aristotle distinguishes between two kinds of virtue—virtues of thought and virtues of character (Aristotle 1999, p. 1103a5–11). The distinction corresponds to parts of the soul. Aristotle believed that a part of our psychology is purely intellectual, while another is part-intellectual and part emotional. Although the distinction is easy to undermine, there is some common sense to it. A failure to grasp logic due to a blunt mind is a different kind of failure than a tendency to erupt into angry outbursts. The angry outbursts are matters of character, and Aristotle thinks that only virtues of character are habits (Aristotle 1999, p. 1103a14–15).

[13] Although Hadot casts the ancient philosophical goal as primarily practical—i.e., an *ethos*—he does accept that theory is a part of that practice. See Hadot (2002), p. 175, par. 3. As I said before, I think there are good reasons to reject the primacy of practical reason in philosophy, whether this was the ancient view or not.

Yet I think there are grounds for disagreement. Aristotle was of the mind that only virtues of character properly required habituation, that habit made quasi-rational emotions stay in place, serving reason. Character needed a kind of behavioral reinforcement, whereas intellectual excellence was simply in the mind once learned (Aristotle 1999, p. 1103a14–15). He said the emotions must learn to *obey* (Aristotle 1999, p. 1102b36). Habit is obedience for unruly hearts. But the mind does not need to obey reasons—it needs to see them. Aristotle created a picture whereby we realize truths intellectually, but have to build them into our hearts through habit for the times when they interact with our emotions.[14]

Yet learning how to think does require habituation, just as learning how to be courageous does. It does not take the arduous task of writing a dissertation to see this. Being objective—for instance, by seeking criticism of one's own position—is an achievement of mental habits just as even-temperedness in the midst of political life is. I discipline myself to analyze arguments and consider objections using much the same *kind* of method—a practice—as I do when I discipline myself to stay level headed around disrespectful people. And the practice results in habits: look for reasons; detach oneself from the goad of impulsive reactions in one's heart. On this one point, Aristotle's picture of the intellect is too disembodied and unemotional.[15]

Nonetheless, it seems unobjectionable that virtue be seen as a habit, although I will not be able to argue the matter fully here. People of good character simply act well without thought except in exceptional circumstances provoking deliberation. That suggests they act well out of habit. As Bernard Williams famously quipped, stopping to think about what to do in most cases is one thought too many for good character. Steal? Not steal? Virtuous people don't steal as a matter of habit. Unless the poor are starving and the rich man's granary is full, you don't think about it (Williams 1981).[16]

Suppose, then, that virtue requires habit. Thus the pursuit of virtue itself is inextricable from the pursuit of virtuous habits. And that means that any ethics of virtue must be concerned with forming habits.[17]

[14] But notice how Aristotle himself throws the division between mind and heart into question at 1144a29–1144b1 and 1144b31–1144b32. The virtue of intelligence as well as wise judgment both require habits of the heart.

[15] Consider Dewey (1916, p. 142) on this point, remembering Greek gymnastics: "It may seriously be asserted that a chief cause for the remarkable achievements of Greek education was that it was never misled by false notions into an attempted separation of mind and body." Here, there is a gymnastic insight that a trained mind is habituated like a trained body. In any event, Aristotle's distinction between kinds of virtue seems to mirror the distinction between theoretical and practical reason, at least in some way. Currently, I am unclear how, since both kinds of virtues are after the good (that is, inside practical reason). Also, I have the reasons to question the distinction mentioned above.

[16] I learned this expression of Williams in the classes of Candace Vogler at University of Chicago in the 1990s. I am adapting it from a different context, which was originally Williams's critique of consequentialism.

[17] Paradoxically, one effect of the situationism debate in contemporary, analytic, virtue ethics is to turn attention to habits, because *these are what is missing* when situations solicit vicious behavior. See K. Anthony Appiah (2008), esp. Chap. 2, and see the index for "situationism", p. 272. For the most troubling situation to solicit vice I have found, and the turn to heroes as guides to habit, see Susan Neiman (2008, pp. 335ff) which has implications for Abu Ghraib. What is paradoxical about the situationism debate's effect is that proponents of situationism wish to show how virtue ethics is useless, whereas I believe they

Happily, the ancients understood these conclusions. Ancient philosophical schools aimed to train the whole character of people.[18] According to Pierre Hadot, the primary method was "ascetics"—exercises designed specifically to develop virtuous habits in someone (Hadot 2002, pp. 189–190). Wisdom came in the *form* of these "spiritual" exercises, which ranged from gym to dialectical drills, and had at their center such things as the examination of conscience (Hadot 2002, *passim*). I will illustrate the examination of conscience shortly, although its most famous forms can be found in Seneca's *Letters* and Descartes's *Meditations* (themselves modeled on Seneca's *Letters*?) (Hadot 2002, pp. 264–265).

Suppose, then, that we want to develop environmental virtue itself. It will take more than a theory of virtue.[19] We will need a practice of virtue, training.[20] And if we wish to write on virtue itself, we will need a way to put that practice into writing. How does one write so that virtue becomes a habit? Here, how one writes about virtue implies something beyond theory. In fact, it implies doing something other than writing *about* virtue—as if virtue were a theoretical object an arm's distance away. Rather, it implies becoming virtuous *in the writing*. It implies writing in such a way that a reader begins to acquire virtuous habits while thinking about virtuous ideas.

Ancient philosophy had a tradition of writing in a way that accomplished the complex, formal goals of philosophy suggested here. The purpose of this section and the last has been to set up an example of such a form of writing, a spiritual exercise, specifically, what was called "the examination of conscience" (Hadot 2002, pp. 198–202).[21] I would like now to turn to the example. It addresses an environmental problem of massive proportions and significance. Here, I take virtue ethically, and not merely theoretically.

Footnote 17 continued
have only shown why Aristotle focused on habits and role models (e.g., the man of wise judgment from book VI of the *Nicomachean ethics*).

[18] Interestingly, some Jesuit schools still have this goal. Is it a coincidence that Ignatius of Loyola (2007) popularized the expression "spiritual exercise" and made spiritual exercises central to Jesuit practice?

[19] For instance, the mainstream, academic interpretation of Ronald Sandler (2007). I believe, though, that one can read Sandler's book as an ethical practice. After all, Sandler shapes his book so that it creates a fine-tuned response to a specific problem for practical judgment, that of genetically modified crops. The book ends with Sandler's application of his theory to selecting which GMC's are worth promoting and which are not. That looks like the search for virtue, not simply the search for knowledge of virtue. The book aims at a practical judgment and enacts it.

[20] As in Henry David Thoreau (2000) and Rachel Carson (2007), or Jeremy Bendik-Keymer (2006, Chap. 9). We will need to see why Nietzsche, one of the profoundest readers of the Greeks, wrote the way he did.

[21] Hadot quotes Seneca: "What a good sleep follows the examination of one's own self! How tranquil, deep, and free it is, when the mind has been praised or warned, and has become the observer and secret judge of its own morals!" (Seneca 1995; in Latin: III, 36, 1–3).

An Example of Form Matching Content

> What is offensive about [our exceeding the natural rates of extinction by fiftyfold] is ... the maelstrom of killing and insensitivity to forms of life and the sources producing them.
> —Holmes Rolston III (1988), *Environmental ethics*, p. 133

Normally, I am a conscientious man. I do my duties almost all of the time. Doing the right thing matters. When I hurt someone, it bothers me, and I am sorry. I try to make amends. If I realize my habits are bad, I try to change them. All these things and more seem typical and make me no different than the average person, and in any event acting as every human should.[22]

There have been times I have been confused by what to do when I tried to do the right thing. Usually, two duties pulled against each other. For instance, I would want to protect my friend from pain, but also knew I had to respect his autonomy. These kinds of tensions are a normal part of moral life, and people deal with them often.

However, right now I am faced with a confusion of a different kind. I do not know which duties I have at all.[23] Yet my conscience troubles me, and has for some time now, never becoming clearer and never subsiding. I feel I have a horsefly buzzing around me, waking me up when I drift to sleep. But I feel my life is like a large workhorse, and the fly is so small that while it keeps me awake, it also does not tear me away from my work and make me address what bothers me.

Therefore today, I have decided to take time to deal with my conscience. I have time. The church bells toll noon on this weekend, and a fan keeps me cool as I think and as I write. I am going to face my confusion and see if I can make some progress in figuring out what I am supposed to do, or if—by contrast—I have simply been the victim of an oversensitivity of some kind and have nothing I am supposed to do. In either case, I will be able to quiet my mind.[24]

So, then, let me take this time now, turn off my cell phone, refuse to open my internet and to read mail, and let me refuse any distraction I might enjoy on this first day of the weekend after a busy week. No music, no trips to the mall to see movies, not even shopping for food I will need. The run I would like to take will wait until later. The book of essays will be read another day. Let me take these few hours to do something far more vital to my soul than any of these things.[25]

* * *

[22] This spiritual exercise begins with an everyday person of sound moral character.

[23] Here, one thought extra *is* called for. Virtue has broken down, or is inexact.

[24] This attention to something you have to *do* is typical of a spiritual exercise and is something self-help manuals have over academic theory, at least from the standpoint of learning virtue itself.

[25] An anonymous reviewer for this journal commented on the narrative dimension of this section and referred to the body of literature—e.g., Martha Nussbaum (1992) being the *locus classicus*—on how narration figures importantly in teaching virtue. But as I will discuss in this paper's last section, my emphasis here is not on teaching virtue, but on *learning* it, and the exercise's point is in the *writing,* not the reading. In other words, this section does not emphasize narration for the reader so much as action on behalf of the writer.

What troubles me came in snippets. Over the Internet, in journals I sometimes read, or which colleagues pass on to me. It appeared in short conversations with people who are better informed than I. Recently, for instance, I was reading a book meant for new college students. The author wrote:

> Many biologists believe that the sixth major wave of extinction since life began is now occurring, and that this one, unlike the other five, is being caused by human action (Jamieson 2008, p. 6).[26]

What has been on my mind is "mass extinction."[27] I've heard it referred to technically as "anthropogenic mass species extinction," or a global wave of extinction caused by humans. It sounds ominous, yet also vague. Is 1% species extinction massive, or 5%?[28]

I did some research and found that there is a range of estimates for the number of species that will go extinct this century. Neither end of the range leaves me feeling comfortable. The worst estimates say the planet will become "like mars" with only 200 million *humans* left on it (Lovelock 2006; McKibben 2006). In such a scenario, most of life on Earth will have been wiped out as a result of many things, but mostly as a result of global warming.

A more moderate estimate by a celebrated biologist at my nation's oldest and most renowned university suggests 25% species loss over this century (Wilson 1999). Then there is a much-cited United Nations report from 2002 predicting that 25% of the world's mammals will be extinct by 2032 (Podger 2002). And there are dire things to say about other kinds of species, too. One report discussed in *National Geographic* put 1,000,000 kinds of land plants and animals extinct by 2050 due to global warming (Roach 2004). Finally, on a more cautious note, I have heard of 20% of all species at *increased risk* of extinction over this century if the globe warms as it has over the past 50 years (I.P.C.C. 2007).[29]

All of these figures trouble me. The respected biologist thinks that we will lose at least a quarter of all life forms throughout this century. That is staggering. All I have to do is go outside, or—right now—look through my window, and imagine that out of every four varieties of plants, insects, birds, and the occasional mammal I see,

[26] For accessible resources on the "current mass extinction," see Ulansey 2008. For a simple route into the phenomenon, see MNBC's website on global warming and species extinction (MNBC 2007). And Al Gore (2006, p. 163) claims our extinction rate is "1,000" times above average.

[27] David Ulansey (2008), who runs the "current mass extinction" site, writes that a 1998 article alerted him to the problem (Warrick 1998). An earlier text than the article that disturbed Ulansey is Richard Leakey and Roger Lewin (1995).

[28] Honesty is important in a spiritual exercise, even if it betrays ignorance or lack of expertise. This is how most people are, i.e., most of us are not experts in many matters.

Also, attention to the lived reality of everyday life is important to bring to mind, since when we aim to *live* virtuously, we have to see how our virtue or lack thereof come up in the mundane details of our lives.

[29] The IPCC estimate is for 1.5–2.5°C temperature rise and places species at "increased risk" of extinction. Note that the warming rate per decade for the last 50 years is 0.13°C, and that the projected range for warming is 1.5–4°C, higher on the upper end than the IPCC's range. See the National Oceanic and Atmospheric Administration's (2008) National Climactic Data Center, questions 3 and 7. Since writing this spiritual exercise, the International Union for the Conservation of Nature (the IUCN) has verified that the 2002 UN report on mammal loss has been correct in its predictions so far and that as of 2008, 33% of the world's amphibians are extinct or at risk of extinction. See James Kanter (2008).

one will no longer be found on Earth by the time the grandchildren of my grandchildren see the light of day. Certainly, as I have read, many of the species that will go extinct include odd forms of life I never see: lichens and mosses, and even things no human eye can discern. But that does not assuage my conscience, as I find all these beings fascinating and part of life, the only life in the universe as far as we know.

How strange the day seems when I think of these reports, looking out my window and erasing with my mind's eye one out of every four species. As I look into the thick, green texture of summer here in upstate New York, instead of the four kinds of leaves I see, there might be only three found on this planet at century's end. Which one goes? All of them are pleasing to the eye and fascinating to look at more closely. And they make all the more interest overlapping each other on this lush, summer day. If ever I have children, I want my children to see them, and we can study them together and make comparisons.[30]

At the same time, I know that most species loss occurs in what people call "biodiversity hot spots"—areas such as coral reefs or tropical rainforests where the diversity of life is dense and lush. I also know that a good portion of estimated species loss comes from loss of insects, fungae, amphibians, and so on—that is, from species to which I have little connection, unlike trees.[31] I am showing my ignorance when I simply subtract species proportionally from my present environment.

I know. But I am trying to imagine something unimaginable to me. Take the songs of birds I hear nearby. There are many calls. One of my favorite is the cooing of doves. Will it go? It is a question of what the dove needs. Global warming throws off the hatching cycles of birds and insects, sometimes making it harder for birds to find the insects they need (Philander 2008, pp. 54–55).[32] Also, without birds and insects, much of life comes to a halt. Plants are not pollinated as readily. And on the back of plants, so much life rests. What will protect the dove, or any of these sounds that remind me I am part of the continuum of life and need not be absorbed simply by human concerns? What will make me human by taking me out of the human?[33]

I think these thoughts, and have heard those occasional items of news, but in general I don't hear people talking about the sixth mass extinction. It's not an item of politics, certainly not in the 2008 presidential election of the United States of America. Yet we are wiping out a truly massive portion of life. There are only

[30] Of course, species extinction will not work as the author depicts it. Central New York may not be affected much at all, whereas biodiversity hotspots in jungles or around coral reefs will be decimated. The author's thought experiment is heuristic: to try to picture the loss of life in a way he can understand. We should remember, too, that species exist *interdependently* in their eco-systems. At a certain point, depleting species can simply cause an eco-system to crash. To put the idea in oversimplified, everyday terms: in ecology, everything is all bound up with everything else.

[31] Thanks to Ned Hettinger for underlining these points with me. See also Wilson (1999) and Rolston III (1988, Chap. 4).

[32] Note, however, that hatching patterns must be studied with respect to each species and its prey. The results are not always bad for the predator.

[33] This inchoate rush of anxiety that rides quickly over theoretical distinctions is typical of how human psychology works. One can't work on it without acknowledging it first. The spiritual exercise will try to make some headway on it.

10,000,000 recorded species, but apparently countless more unrecorded—one environmental scientist told me recently as many as ten times more.[34] Even if we lead a quarter of the *known* species to extinction, that is over two million species gone by our deeds. I do not know how to think about this. I do not know what my duties are. I do not hear people discussing this matter in everyday affairs, and yet I know everything about this disquiets me, and does not stop disquieting me, even when I push it aside. I feel I am in a dream.

* * *

As I've read, the main causes of extinction relate to our consumption: mining and logging, fishing and monoculture, clear-cutting land for cattle grazing, and real estate development. Then there is the burning of fossil fuels and the population explosion of meat animals producing methane, both of which lead to global warming. Global warming, in turn, leads to shifting climate zones and acidic oceans. All these causes fed by our consumption escalate along with our own population and its resultant scramble for land, shelter, fuel, water, and food (Strieker 2002).[35] As we reach 9,000,000,000 by mid-century and continue an industrial form of life, it is not hard to imagine spreading our effects around Earth into almost every nook and cranny, depth and shoal.[36] We currently exist in a destructive cycle whereby we acquire, consume, reproduce, and acquire yet even more.

Yet a good portion of people in many different countries hold life to be sacred, or rare, or respect-worthy, or magnificent, or wonderful. They school their kids in it. To take just three examples that cover over half the world's population, Buddhists, Muslims, and Hindus all agree that non-human life is worthy of reverence. A morality of reverence for life would appear to be justified to a large portion of humanity, and so we could say that the duty—the duty to treat life decently—is known. The idea that we are currently wiping out a large chunk of life on this planet should therefore strike a large number of people as profoundly disturbing.[37]

When I think about these things, I become even more agitated and feel like giving up. The duty is in many people's hearts, but the behavior—the character—does not display it. Morality does not join with life. All sorts of memories come to mind, of when people did wrong things, knew they were doing it, and yet used sophisticated reasoning to deny themselves responsibility. A kind of inky bitterness builds up in my throat, and I want to explain that people are selfish or wicked.[38]

---

[34] As in the case of turning off one's cell phone and taking time out, *talking with others* and *going to find out* about the cause of one's concern is part of the *habit* of virtue. So the mention of it signals the reader to virtue's form of life, not simply its theory. It's important: this kind of signal to the reader that she's not simply reading a text but thinking about how to actually live.

[35] The big cause, however, is global warming. On this cause, see the sources cited in footnote 29 and the paragraphs preceding it.

[36] Thanks to Dan Meior for helping point out the correct population estimate.

[37] That is, where is the *habit*?

[38] It would be easy to state the contradiction simply as a proposition. But getting ourselves to find and acknowledge uncomfortable contradictions is crucial to virtue itself. It takes time, as Socrates—and by extension Plato—saw in his numerous confrontations with human resistance to uncomfortable contradiction.

But I also know, as soon as I calm down—and I must calm down; I know judgment is skewed when I am angry[39]—I know that the average person is conscientious. Indeed, I began from this point. Remembering this reminds me that I should be charitable and look more closely. Given that people are conscientious—and not normally wicked—given that a morality of reverence for life is justified [for I believe that it is (Bendik-Keymer 2006, Chap. 6)], what is the problem here? As a normal person, I am not inherently wicked. And yet I am part of the unintended sixth mass extinction, wiping out millions and maybe millions more of forms of life.[40]

* * *

When I look at species themselves, and hold them up next to my belief that life deserves reverence, I am struck by something. Not species, but *individual lives* are the primary objects of my reverence for life. Let us suppose a living being crossed my path, a being of a kind I'd never seen. Perhaps it is a worm-like being. Yet it has a strange combination of attributes. Whereas its skin is smooth like a worm, it has what appear to be legs, although they are soft and malleable, unlike an insect's legs. In addition, there's a coloration unlike any worm or millipede I have ever seen. I am so surprised, I take a photo of it with my cell phone.

Now it just so happens I have a friend, a scientist, who specializes in biology and knows many friends. He shows the picture I took around to colleagues, and indeed it appears I saw some kind of mutant, what people a 1000 years ago called a *monstrum*, that is, an individual without a species. My question is, does the fact of this mutant being species-less make it any less convincing that I ought to avoid walking on it?

No. It does not, given that I respect life. If anything, its individuality makes this living being more valuable, since it is more than one of a kind. It is one *without* a kind. A rare, living being.

The point is, not the species but the individual receives my reverence for life. As a Dutchman said 400 years ago, every living being has its *conatus*, its drive to preserve itself in its own being (de Spinoza 2005, III p. 6). And *this* is the main object of my reverence for life—this striving individual—not the universal kind—the species—to which most individuals belong in some form or another (Nussbaum 2007, Chap. 5; Bendik-Keymer and Thompson 2008; Bendik-Keymer 2009).

How does reverence for life commit me to respecting species, then? I see that species themselves still matter to me even beyond the individuals that comprise them. Species matter in some secondary sense. My concern for individuals involves something like justice—an attention *to* them—but my concern for species involves

[39] An important theme among the ancients, especially Seneca (1995).

[40] One way my spiritual exercise departs from some ancient spiritual exercises is in the role that theoretical confusion plays in it. I do not assume that our best understanding is clear on the matter of the sixth mass extinction but that only our psychology and our virtue wavers. Rather, I think that our best understanding is unclear, a matter reflected in our psychology and virtue as well. This fact complicates the charge that I am advocating merely better *application* of theory in virtue ethics. I think theory has an important place in virtue itself, provided it is taken ethically, which the present style attempts to illustrate.

something like wisdom, an awareness of something that matters.[41] After all, the world as it is contains countless forms of life. If all of these were to collapse into the same number of individuals but only half the number of forms of life, I would feel a loss, a loss of the diversity of life.

Yet species go extinct as a matter of course. In fact, some species extinction is good. I have heard it is unlikely that human beings would have evolved if the dinosaurs hadn't gone extinct. Their mass extinction was a precondition allowing mammals to flourish and evolve. With the dinosaurs gone, it was much safer to do so. So reverence for *our* life should support species extinction in some cases! The extinction of species makes way for others.

What am I bothered by then? I am bothered by two things. I am bothered by our causing extinction, and I am bothered by the scale of extinction.[42] A third to a half of all life is so massive that it exceeds my ability to visualize it. *We* are doing massive things, things that even on a smaller scale would bother me. Why do they?

* * *

To begin with, there are many arguments that claim species extinction and loss of biodiversity are bad because of what we might lose when losing the opportunity to use species. A famous version of this argument goes: Suppose some species provides the future cure to cancer, but we make it go extinct now. Wouldn't this be bad? Therefore, we should protect species from going extinct in order to protect potential but at present unknown benefits to humankind (Rolston 1988, pp. 127–130).[43]

There are two main problems with this argument, though. The first is that its reasoning could be used against it. Any species *could* be useful to us at a later point in time. Yet protecting a species might block the way for new species to come into their own, and the new species might just as easily help us, even better than the ones we protect. So we should both protect species and let them go extinct to make room for new ones. This is contradictory.

The second problem is that my moral concern emanates from what we are doing. Proposing that extinction itself is bad because of what we might lose doesn't get at the *relation* between our acts and their effects. After all, if species can be useful to us, it doesn't matter whether we put them at risk or something else does. The risk is bad all the same. But I am currently troubled by what *we* are doing. *I* feel like I am doing something wrong.

In fact, the usefulness of species does not figure at all in the central point of my conscience. Yes, I do think that if half of the world's species go extinct, it will hurt us. After all, the United Nations Environmental Program predicts that by mid-century, we will see 150,000,000 people fleeing environmental problems in the

[41] Michael Thompson (2006) calls this a distinction between bipolar and monadic norms, that is, norms involving two agents (relations of right) and norms involving only one agent and objects that matter (practices protecting the good).

[42] Here, you can see that the emphasis is on learning a *process*—the examination of conscience. In effect, what we are learning is a kind of know-how, i.e., practical knowledge, the knowledge of *how* to examine our consciences.

[43] David Keymer helped me formulate the last sentence of this paragraph.

world (University of Iowa Center for Human Rights 2007). These people are called "environmental refugees." A total of 150,000,000 of them is almost half the population of the United States of America. These refugees will suffer, and the political and economic instability they are likely to cause will expand the circle of suffering beyond them. While many of these refugees will flee desertification or flooding, it is not unimaginable that species loss could contribute to the need to flee areas in search of more habitable or nourishing ones. Or that the price of food will cause regimes to collapse, driving people away from them.

But when I look closely into myself, I see that what bothers me is what we are doing to *other* forms of life. I am bothered by the fact that our actions devastate life on Earth. The question is, apart from effects on us, what—if anything—is wrong with devastating life? And in particular, although I do care about living individuals, what is wrong with devastating *forms* of life?

The beetles and bees, the microscropic life in the sea, the many plants and birds, these all create the conditions on which life as a whole depends. Winnowing, thinning out life creates dangerous conditions for much of life, due to the interdependency of forms of life on each other in pollination, predation, decomposition, and so on.

Yet the narrowing of life's diversity *now* leaves room for future diversity, for diversity *then*. Just as we emerged out of the shadow of extinct dinosaurs, so a mass extinction might clear the way for future development, development of new diversity. That is, even if diversity matters, why should *present* diversity matter? After all, *no* predictions suggest we are likely to eradicate *all* life. According to expert estimates, life will go on, mass extinction or not. And to underline the point further, when we switch from human time to geological time, it's clear that all human history is but a second or two on the clock of geological time. What we do to life is insignificant from the standpoint of geological time, unless we destroy the Earth itself.[44] If form gives way to form, destructions should not bother me.

Is it just that I am attached to this world, to *its* species, to what has come to be? Yet that doesn't seem right. After all, I am troubled by the scale of our destruction, and that destruction takes in countless species of which I do not even *know*, much less feel attachment. Secondly, it seems arbitrary to hang the protection of species on what I happen to like, even deeply. What you like, deeply, might be different. Morality does not have this character of being decided by taste. Its reasons are universal, whereas taste is subjective.[45] But if my *moral* disquiet does not issue from my attachments, what is to preclude that a mass extinction now is acceptable, because it can clear the way for future life? According to that view, it might even be *im*moral to resist species extinction. Pulled between the possibility of future diversity and the amoral appeal to attachment, I find it hard to clarify what's bothering me.[46]

[44] Joshua Graae helped me underline the point.

[45] The *locus classicus* for these points about morality and taste is Immanuel Kant (2007) "The Analytic of Aesthetic Judgment," originally published in 1790.

[46] Although I depart from his reasoning significantly, Ben Bradley (2001) helped me think through a number of issues relating to biodiversity. Bradley deals with the work of Holmes Rolston III, Robin Attfield, Nicholas Rescher, and Robert Elliot, among others. See also Rolston III (1985).

* * *

I haven't solved my problem, and I haven't dissolved it either. My conscience still bothers me, although I cannot put my finger on the cause. When I try to turn to the species themselves and find a reason they ought to be part of my reverence for life, counter-examples muddy the stream of my reasoning and prove that I do not have a sound position for revering species just as they are over future ones. I feel that my instinct to protect them comes down to my attachment to them, and yet I also know that my conscience suggests I am doing a *moral* wrong—not simply a denial of my preferences—by participating in the sixth mass extinction. I feel groundless, and yet worry I am being unreal, misguided by abstract reasoning away from what any decent person would see looking at the scale of destruction we are causing even as I write.

What would a decent person see?[47] First, injustice, which concerns individuals (Nussbaum 2007, Chap. 5). Any extinction comprises the deaths of numerous, striving individuals. A mass extinction is mass death. Unless we have good reasons to destroy so many lives, we are already in the wrong (Rachels 2007; Bendik-Keymer 2006, Chap. 8). Yet how would a decent person see massive species loss itself?

Here, I am reminded of a piece of philosophical writing I once read. A famous environmental ethicist spent a good 35 pages working through the question of what, if anything, is morally objectionable in the loss of species themselves.[48] He led me to see that what he called "the speciation process" is indistinguishable from the history of life. He called the process "the wonderland," and what he meant, I think, is that it is the macro-order of life, by which any individual living being comes to be. Seen from his perspective, species loss is a loss in the macro-order of life. It is loss of "living historical form" (Rolston 1988, pp. 131, 135).

Such a perspective on life's macro-order gives me a way to understand decency in this situation, for decent people do not destroy thoughtlessly. They are not what we would call "destructive" people, even if they at times have to take or destroy for well thought-out reasons. I think the decent person *would* approach the loss of species themselves through the prism of order, because that would allow her to grasp the way life is at stake in species loss, even though she would not be focusing just on individuals.

Species form the historical and ecological order of the world as we've inherited it, and I have an intuitive respect for order, unless good reasons apply to breaking it. Moreover, living order is an especially wonderful thing. Given that species are the

[47] The appeal to the decent person is a classic move of virtue ethics, at least its Aristotelian form. It is not question-begging if we realize that practical wisdom is primarily know-how (and also what I call knowing others—the form of knowing in relational reason which is key to the social dimension of ethics). When we ask for theoretical clarity, as the questioner is doing here, we might learn by explicating the know-how of someone who just strikes us as virtuous. Interestingly, there is a space of learning that can arise where propositional knowledge (knowledge that, the form of theoretical knowledge) has not caught up with know how (the form of practical knowledge), and this is what I believe appeal to the decent person tries to keep in view. Obviously, a full defense of this kind of move would take a paper on its own.

[48] Here, theory is *taken* ethically. Being such, it is no longer pointless regarding virtue itself. A spiritual exercise should show people *how* to take theory ethically.

common medium of life's functioning and the archive of its history, the extinction of species is not simply a reduction of life as we know it, as if we were turning down its volume. It is the destruction of life as we've inherited it, the distortion of its worked-out song. Rachel Carson (2002), the great environmentalist, understood this when she warned against a "silent Spring"—a life where birds no longer exist due to pesticides. When we lose many forms of life, we lose also their interrelationships, which make up the order we've inherited and on which many forms of life depend. We lose the efforts of living history that have profoundly shaped who we are (Snyder 1990). In other words, we destroy the *quality* of life, not just its quantity.[49]

As I have said, there may be reasons to think that the species themselves we render extinct might be balanced in some sense by future species that will evolve, that a new kind of sound will coalesce a million years from now. Yet we are not presently considering this trade-off and the intervening silence. We are not discussing the loss of history, losses in the macro-order of life. We risk dead seas, deserts, and silent Springs—and a gutted, looted museum of the history of Earth.[50] A decent person would therefore see that we are chaotic. We thoughtlessly destroy an order we are lucky to inhabit, and do so without a whimper.[51]

"To destroy." What does that mean? It has taken millions of years for the world of life as we know it to evolve. Its forms are survivors of chance, "momentary cosmic accidents."[52] We hardly understand how they fit together. Yet they are forms beyond anything we have ever created. This makes them worthy objects of reverence. I read about reverence recently. A popular philosopher wrote:

> Reverence is the feeling you have for something none of us will ever reach. ... Reverence is awareness of human limits.... [Y]ou're in the grip of something vaster than you are (Neiman 2008, pp. 232–235).

For all practical purposes, that describes an attitude that someone with reverence for life does have: awe and humility before the complexity of life, its order. That this order has come about by chance is even more reason to revere it, for chance is fully out of our control. It seems that a decent person would at the least find our hurtling thoughtlessly into the sixth mass extinction wrong, *because it is irreverent*.

---

[49] See also Rolston III (1985, p. 722): "It is not *form* (species) as mere morphology, but the *formative* (speciating) process that humans ought to preserve." See, finally, Wilson's (1999) opening section. He emphasizes the destruction of life as we know it, whereas Rolston III emphasizes the destruction of life's history.

[50] The use of strong metaphors is typical to spiritual exercises among the ancients, because of the way intense images can capture and train the will. One ancient master of this technique was Plutarch (1992). See, for instance, his essay on *euthymia*—"contentment."

[51] On luck, see Gould (1996). On the whimper, see Eliot (1952), originally published in 1925. See also Rolston III (1985, p. 720): "What is offensive in the impending extinctions is ... the maelstrom of killing and insensitivity to forms of life and the forces producing them."

[52] Stephen J. Gould (1996, pp. 4 and 18) speaks of "the unpredictability and contingency of any particular event in evolution and emphasizes that the origin of *Homo sapiens* must be viewed as, an unrepeatable particular... ." He claims our life is "in union with other creatures as one contingent element of a much larger history," and that "we are, whatever our glories and accomplishments, a momentary cosmic accident that would never arise again" even if we repeated the conditions that gave rise to us.

And that makes sense. My conscience is all disquieted, and an Enlightenment philosopher once said that conscience is the love of order.[53] Yet our irreverence tears order apart, hard-won, lucky order.

Here, let me visualize the loss. I will make a list in my mind of the dynamic forms this order takes. I won't make a poem, but simply list the expressions of order slowly, taking my time to visualize each expression before moving on to the next.[54]

We lose order that:

feels
stirs
moves
breathes
swims
flies
grows
spreads
reproduces
blooms
drifts
evolves.

Sure, reverence for life focuses on individuals. But I now see that life's chance order can be its object, too. That's what I felt was linked to wisdom, to being aware of what matters.

To be precise, irreverence comes down to thought*less*ness. We can see this if we consider the case of having good reasons to destroy order, of which there are many. If a species led to the death of all human life, there would be good reason to render it extinct. But there is a big difference between this kind of justified, deliberate destruction and wild, thoughtless destruction. Artists, for instance, contest form thoughtfully. Their destruction does create the space for new form.[55] But the sixth mass extinction is not thoughtful in this sense or in any sense. At large in the public sphere, there is no serious discussion of what we are doing, and there is very little awareness of the extinction itself. The point is precise: thoughtless, *without a thought*.

Especially since the industrial revolution began, in the space of only 200 years, we have managed to set in motion a massive rending of our world's lucky form.[56] Even if I can't argue that each living species today has a reason it must be preserved, and even if chance will absorb our destructiveness in its branching

[53] Jean-Jacques Rousseau (1969, p. 936) claims conscience is the love of order. Apparently, his view harkens back to Malebranche. See also Bendik-Keymer (2002, Chap. 2).

[54] Note again the emphasis on a technique of visualization, on something we *do*.

[55] Example, Einstürzende Neubauten's (1981) music from the 1980s.

[56] Some will argue that increased anthropogenic species extinction rates can be traced back thousands of years to the technological conditions that allowed our population to increase, our land use to magnify, and our hunting and fishing to become more efficient. I concede this point, but still note that the scale and rate of species loss jumped massively in the twentieth century as our population exploded and does so currently in the face of climate change.

proliferation of life, the *way* we are acting legitimately bothers me. It is shocking behavior, the kind I would never permit my children, my students, or myself. Too much is at stake, we know too little about what we do, and we have not stopped to think.[57]

The *manner* of our action, the way we are heading into the sixth mass extinction is the problem. The thoughtlessness of how we act displays the vice of irreverence. And the vice of irreverence overlooks the quality of life.

I said at the beginning of my reflections this morning that most people are conscientious, and I conceded that most of us have reverence for life. Yet I now see how the problem is deeper. We are acting chaotically. There's a disconnection between our moral convictions and the outward vice of our behavior. Responsible people do not act this way. They don't believe one thing in their hearts and then contradict it.[58]

Yet if the problem is a failure of responsibility, the problem is located *behind* conscience, before we ever listen to it, or think that we do. For we have, first, to be in the habit of stopping and thinking for conscience to guide us. We have to be in the habit of *listening* to conscience. Irresponsibility undercuts both habits. It's not in a *position* for conscience. And when it comes to the sixth mass extinction, we show we're in no such position. We have not stopped to think, not as a people, and nowhere on Earth where I've lived, which is on several continents. I am beginning to revise my view that we are conscientious, because we do not appear to be in a position[59] to be conscientious.

Where, then, is the anxious chaos spinning against our hearts? There is a disconnect between our hearts and our manners, between morality and life, between the lofty realms of our consciences and the base realms of practical, everyday life. Why? Thinking about it, I think the root disconnect is between effect and cause. The effect is mass extinction, and we are the cause. But we are not owning up to the effect, taking it as our own. It is like doing something, but acting as if we were not

[57] Focus on our *manner* of action is a mark of virtue ethics.

[58] Here, what—if anything—is novel about my form of spiritual exercise opens up. Classic spiritual exercises come to philosophical closure at their end—or at least purport to (think of Descartes (1996) as the voice in early modernity for the Senecan tradition). However, my spiritual exercise ends in an *aporia*, a deliberately discomforting gap that should take a reflective person outward to different *kinds* of considerations than the moral ones addressed here. This discomfort is important for virtue itself, because much in our lives cannot be encapsulated solely by the moral. We need to bring in the political, the economic, etc.

[59] This point in the examination of conscience should make readers begin to wonder if there are *extra-*moral reasons for our irresponsibility, and this has been the case in the two audiences to whom this exercise has so far been read. Perhaps our *social* structure—combining political and economic institutions and the way they shape us—has a lot to do with why we are in no *position* to be responsible. The intent of this exercise is *precisely* to take the reader to this *aporia*, thereby to motivate a further spiritual exercise examining social structure—away from the personal, to the institutional. As I've said, in this respect, the examination of conscience modeled here is somewhat novel for its tradition. Traditionally, the examination of conscience comes *back* to the self, closing up the problem within it. But this examination of conscience, while it attempts to focus the self, opens up beyond it in the suggestion that *there is more than the self* to fault in the failure of responsibility. The exercise does not remain closed, but is incomplete. Compare, by contrast, Descartes (1996), which, once locating the problem of falsity in the will, closes up the uncertainty of life in the *self*'s newfound responsibility to judge with only clear and distinct ideas.

acting. Though the rumors of our effects have been in major media sources now for a decade, we do not hear them *and then go to confirm or to disconfirm them.* Practically speaking, *they do not matter to us*, for as the Holocaust survivor Vladimir Jankélevitch once wrote "Do not listen to what they say—look at what they do!" (Jankélevitch and Hobart 1996). Alerted to the effects of our actions, we do *not* bother to find out if the rumors are true.[60] This is where the root thoughtlessness lies: a disowning and lack of consideration of consequence, a kind of autopilot without assessing what we do.[61]

Yet a mature, human being has the habit of checking what she's doing. Maybe it is because I do check what I do, and sometimes check too much, that I was in a position to hear even what little conscience I did. But I am not checking enough, and this implicates me in irreverence, an irreverence I do not support. Unlike today, unlike an adult, I acted thoughtlessly, destroying the order in which we live, without a good reason to do so, without even a thought about looking for a reason or a minute to spare to find out if the rumors are true.[62]

Remarks on Spiritual Exercise

Even though I have claimed spiritual exercise fits the nature of the philosophy of virtue, one objection to it in a forum such as our own is that the exercise seems better intended for non-academics, rather than scholars. After all, we (theorists) have thought a lot about the issues of this exercise and their significance. Those who have not, by contrast, are non-scholars, many everyday folk who don't have the time to stop and think. An exercise like the one just given seems suited to them, and ill-suited to academics. It seems to be a kind of rhetoric for people who usually don't think. We, by contrast, need clear knowledge about what we are supposed to think. We need good theory.

But, even though non-academics *do* think (and often better than academics when it comes to practical life!), the exercise is not rhetoric. Its goal is virtue itself, theory taken ethically. Furthermore, it is not suited only for non-academics, for we (theorists) should not be too sure we actually are aiming at virtue itself. After all, we saw that the life of theory does not imply the philosophical life and that virtue itself is a matter primarily of know-how (in practical reason), not "knowledge that" (in

[60] Though unintended mass species extinction is *not* comparable to genocide (which is intentional), and so should not be compared to ecocide (which is likewise intentional), the dullness with which we respond to the rumors of it in the media bears comparison to the kind of human inertia and weakness of will that has plagued reports of genocide. I guess we can say: If people won't respond to genocide, why think they would respond to unintended mass species extinction? On the suggestion that we are committing ecocide, see Roger Gottlieb (2003), and on the claim that we are, see Frederick L. Bender (2003).

[61] The thoughtlessness I have acknowledged is thus much simpler than the kind Martin Heidegger discusses in his many writings (cf. Heidegger 1982). Rather, the inspiration for this essay began by thinking about what Hanna Arendt (1994) meant by "thoughtlessness." I sometimes think that our thoughtlessness is a form of evil's banality, although I am unsure.

[62] In Ronald Sandler's (2007) terms, I realize that I lack the *dispositions* of a virtuous agent, in particular, an agent who expresses the virtue of reverence for life. Rather, my dispositions are vicious—a *lack* of reverence for life *in practice*.

theoretical reason). The purpose of the exercise is to awaken people—theorists and non-theorists alike—to virtue itself, specifically concerning the utterly disquieting prospect of the sixth mass extinction.

So how might we distinguish a genuinely practical dimension of the exercise from a merely theoretical one, given that the exercise does trade in abstract conceptualization? The answer is through the form of *questioning*. A questioning that seeks know-how is different from a questioning that seeks only knowledge. And the exercise's form of questioning displays this difference. In the exercise, *the living person* thinks, rather than simply the intellect.

Consider. When it comes to building virtuous habits, habits of conscience, the important thing is not just to make a person *think*, but to make sure it is the *person* who thinks. What I mean by this is not hard to grasp once you think about it. There is a world of difference between thinking about a detached intellectual problem that does not emanate from our hearts and an intellectual problem growing from heartfelt disquiet.[63] The exercise, however, comes from, and always orbits around, heartfelt disquiet. *This is its form of questioning*. In so doing, the exercise makes sure any reflection brings in the whole person's existence,[64] rather than just some part of it that can easily be compartmentalized as an intellectual pursuit and forgotten. But by bringing in the whole person's existence, the reader is *already* seeking practical wisdom, whereas a paper appealing just to our intellects does not imply seeking any kind of knowledge of how to live. *The existential connection is key.*[65]

This practical dimension of the exercise—what I am calling its "existential connection" in the form of questioning—allows us to make more than one helpful distinction. First, it allows us to emphasize that even though theory can have a practical *effect* on a person, it is another thing to take theory in such a way that is has a practical *point*. That is what taking theory ethically—thinking out of an existential connection—does. It makes sure that the point of theory is living, which is a requisite for using knowledge to become more ethical. Secondly, the practical dimension of the exercise allows us to understand that the point of the exercise is in the *writing*, not the reading. The ancients shared their spiritual exercises to spur each other to do their own, or—in Plutarch's case—to keep their own "notebooks." A reader *could* take most anything ethically, but in making the distinctions I have, I have been focused on the writing and its point. It is the writing that aims at virtue

[63] One might object that a neurotic theoretician can feel heartfelt disquiet at a purely intellectual problem with no clear relation to wisdom. True, but I would say she is approaching the problem philosophically, although making a mistake. The heart is the seat of the emotion love, and love's object is a good. To have heartfelt disquiet driving one's questioning is therefore to question while engaged in a search for something that relates to the good. This is the province of seeking wisdom, however, inchoately.

[64] An existence that seeks some view of the good. See Aristotle (1999, pp. 1095b15–1095b17).

[65] This is a point Kierkegaard understood well (Kierkegaard 1988, 1983), making his writing far more fitting for virtue ethics than most virtue theory. He called his style "upbuilding." See especially *Either/or* (1843), vol. 2, "A final ultimatum," where the style was first attempted. Also, note especially his invocation of the heart's disquiet in his most mature and elegant work, esp. *The Sickness unto Death* (1848), preface. Kierkegaard learned to write either from disquiet or from love. In either case, he wrote from conscience. Whatever the merits of his theology, his understanding of the *point* of virtue ethics, like Nietzsche's, is unparalleled today.

itself and proves that aim in the way its point manifests itself in the form of questioning. Thus, a spiritual exercise is a different *kind* of thing than, say, a science textbook taken ethically by a reader or that same textbook happening to have an unintended ethical effect on a reader.

The upshot is this: learning virtue itself is primarily a practical matter, as all ethics are. If what we want is to learn virtue, then we must make sure our philosophizing coheres with the ontology of the situation. Being ethical is a different kind of thing than being scientific. To reflect this conceptual logic—what I call "the grammar of being"—we need to adopt forms that fit our content. We can't coherently speak *about* virtue itself and claim to be learning virtue. Rather, we must *pursue* virtue itself.

At the end, the kind of exercise I have presented has a legitimate claim to be philosophical because its form and content cohere. The discourse is about virtue and is written *in such a way* as to ignite and train virtue. People are not just reading about passion and how it must be shaped in their lives; they are shaping their passion in their lives. Yes, the exercise is preliminary and meant to be copied, but through it, people are not just theorizing about how thoughtlessness is at the root of the sixth mass extinction; they are confronting their own thoughtlessness and discovering an *aporia*.

Acknowledgments Thanks to many people who helped with this paper, but in particular, to Ken Baynes, Ben Bradley, Tom Brockelman, Phil Cafaro, Amber Carpenter, Aaron Conte, Eva Fernandes, Tabor Fisher, Joshua Graae, Kevin Hayes, David Keymer, Mari-Ann Kucharek, Irene Liu, Dan Meior, Sherine M. Najjar, Ron Sandler, Lauren Tillinghast; participants at the conference *Human Flourishing & Restoration in the Age of Global Warming*, Clemson University, September 5th, 2008; colleagues at the Le Moyne faculty symposium; three anonymous reviewers for this journal who did their job well; the Department of Philosophy at Case Western Reserve University; Irad Kimhi; and Elaine Wolf Steinberg, who surprised me at the end of the writing process. The long example of a spiritual exercise in this essay is the first of approximately six spiritual exercises comprising a short book-in-progress, *The Sixth Mass Extinction: Spiritual Exercises for the Sake of Life*. Each exercise takes the reader through a different dimension of the ethical, political, and economic problems involved in the sixth mass extinction. The link between each is an *aporia* ending a particular exercise and leading one to take up the next.

References

Anthony, K. A. (2008). *Experiments in ethics*. Cambridge, MA: Harvard.

Arendt, H. (1994). *Eichmann in Jerusalem: A report on the banality of evil*. New York: Penguin Classics.

Aristotle. (1999). *Nicomachean Ethics* (T. Irwin, Trans.). Indianapolis: Hackett.

Bender, F. L. (2003). *The culture of extinction: Toward a philosophy of deep ecology*. Boulder, CO: Humanity Books.

Bendik-Keymer, J. (2002). *Conscience and humanity*. Dissertation submitted to the University of Chicago Department of Philosophy, U.M.I./ProQuest, Ann Arbor.

Bendik-Keymer, J. (2006). *The ecological life: Discovering citizenship and a sense of humanity*. Lanham, MD: Rowman & Littlefield.

Bendik-Keymer, J. (2008). Review of Pierre Hadot, *What is ancient philosophy? Philosophical Practice, 3*(2), 302–303.

Bendik-Keymer, J. (2009). From humans to all of life: Nussbaum's transformation of dignity. In F. Comim & M. Nussbaum (Eds.), *Capabilities, gender, equality: Toward fundamental entitlements*. New York: Cambridge University Press.

Bendik-Keymer, J. & Thompson, A. (2008). Injustice & species extinction. In Eco-Res Forum (Ed.), *Environmental (in)justice: Sources, symptoms and solutions.* 11–24 April. http://eco-res.org/phpBB2/viewtopic.php?t=68. Accessed 22 July 2008.
Bradley, B. (2001). The value of endangered species. *The Journal of Value Inquiry, 35*, 43–58.
Carson, R. (2002). *Silent spring.* New York: Mariner Books.
Carson, R. (2007). *Under the sea wind.* New York: Penguin Classics.
de Spinoza, B. (2005). *Ethics* (E. Curley, Trans.). New York: Penguin.
Descartes, R. (1996). *Meditations on first philosophy* (J. Cottingham, Trans.). New York: Cambridge.
Dewey, J. (1916). *Democracy and education.* New York: Free Press.
Einstürzende Neubauten. (1981). *Kollaps.* Berlin: ZickZack.
Eliot, T. S. (1952). The hollow men. In T. S. Eliot (Ed.), *Complete poems and plays, 1909–1950.* New York: Harcourt.
Gore, A. (2006). *An inconvenient truth: The planetary emergency of global warming and what we can do about it.* Emmaus, PA: Rodale.
Gottlieb, R. (2003). *A spirituality of resistance: Finding a peaceful heart and protecting the earth.* Lanham, MD: Rowman & Littlefield.
Gould, S. J. (1996). *Full house: The spread of excellence from Plato to Darwin.* New York: Three Rivers.
Hadot, P. (1998). *The inner citadel: The meditations of Marcus Aurelius* (M. Chase, Trans.). Cambridge, MA: Harvard University Press.
Hadot, P. (2002). Philosophy and philosophical discourse. In *What is ancient philosophy?* (M. Chase, Trans.), (pp. 172–233). Cambridge, MA: The Belknap Press.
Hadot, P. (2003). *Philosophy as a way of life: Spiritual exercises from socrates to foucault.* A. Davidson (Ed.). New York: Blackwell.
Heidegger, M. (1982). *The question concerning technology, and other essays* (W. Lovitt, Trans.). New York: Harper Perennial.
IPCC. (2007). Summary for policy makers. In *Fourth Assessment Report: Working Group II Report 'Impacts, Adaptation, and Vulnerability'.* http://www.ipcc.ch/ipccreports/ar4-wg2.htm. Accessed 25 November 2008.
Jamieson, D. (2008). *Ethics and the environment.* New York: Cambridge University Press.
Jankélevitch, V., & Hobart, A. (1996). Do not listen to what they say, look at what they do. *Critical Inquiry, 22*(3), 549–551.
Kant, I. (2007). *Critique of judgment* (J. C. Meredith, Trans.). New York: Oxford.
Kanter, J. (2008). One in four mammals threatened with extinction, group finds. In *The New York Times.* 7 October. Accessed 8 October 2008.
Kierkegaard, S. (1983). *The sickness unto death* (H. Hong & E. Hong, Trans.). Princeton: Princeton.
Kierkegaard, S. (1988). *Either/Or* (H. Hong & E. Hong, Trans.). Princeton: Princeton.
Leakey, R., & Lewin, R. (1995). *The sixth extinction.* New York: Doubleday.
Lovelock, J. (2006). *The revenge of Gaia: Earth's climate in crisis and the fate of humanity.* New York: Basic Books.
McKibben, B. (2006). How close to catastrophe? In *The New York Review of Books*, November 16th. http://www.nybooks.com/articles/19596. Accessed 9 July 2008.
MNBC. (2007). *Climate change is changing species.* http://www.msnbc.msn.com/id/17861866/. Accessed 9 July 2008.
National Oceanic and Atmospheric Administration. (2008). *Global warming: Frequently asked questions.* http://lwf.ncdc.noaa.gov/oa/climate/globalwarming.html. Accessed 25 November 2008.
Neiman, S. (2008). *Moral clarity: A guide for grown-up idealists.* New York: Harcourt.
Nussbaum, M. (1992). *Love's knowledge: Essays on philosophy and literature.* New York: Oxford.
Nussbaum, M. (2007). *The frontiers of justice: Disability, nationality, species membership.* Cambridge, MA: Harvard.
Philander, S. G. (Ed.). (2008). *The encyclopedia of global warming and climate change.* Thousand Oaks, CA: Sage.
Plutarch. (1992). On contentment. In *Essays* (R. Waterfield, Trans.), 211–238. New York: Penguin Classics.
Podger, C. (2002). Quarter of mammals 'face extinction'. On *BBC News World Edition.* 21 May. http://news.bbc.co.uk/2/hi/science/nature/2000325.stm. Accessed 9 July, 2008.
Rachels, J. (2007). *The legacy of socrates: Essays in moral philosophy.* S. Rachels (Ed.). New York: Columbia.

Roach, J. (2004). By 2050, warming to doom million species, study says. In *National Geographic News*, 12 July. http://news.nationalgeographic.com/news/2004/01/0107_040107_extinction.html. Accessed 9 July 2008.

Rolston, H., I. I. I. (1985). Duties to endangered species. *BioScience, 35*(11), 718–726.

Rolston, H., I. I. I. (1988). *Environmental ethics: Duties to and values in the natural world*. Philadelphia: Temple University Press.

Rousseau, J.-J. (1969). Lettre à Beaumont. In *Oeuvres Complètes*, IV. Paris: Pléiade.

Sandler, R. (2007). *Character and environment: A virtue-oriented approach to environmental ethics*. New York: Columbia.

Seneca. (1995). On anger. In *Moral and political essays* (J. Cooper, Trans.), (pp. 1–116). New York: Cambridge.

Snyder, G. (1990). The place, the region, and the commons". In *The practice of the wild*. San Francisco: North Point.

St. Ignatius Loyola (2007) *The spiritual exercises of St. Ignatius of Loyola* (Father Elder Mullan, Trans.). New York: Cosimo Classics.

Strieker, G. (2002). Scientists agree world faces mass extinction. *CNN*. 23 August. http://edition.cnn.com/2002/TECH/science/08/23/green.century.mass.extinction/index.html. Accessed 9 July 2008.

Thompson, M. (2006). What is it to wrong someone? A puzzle about justice. In R. J. Wallace, P. Pettit, S. Scheffler, & M. Smith (Eds.), *Reason and value: Themes in the moral philosophy of Joseph Raz* (pp. 333–384). New York: Oxford.

Thoreau, H. D. (2000). *Walden, with Ralph Waldo Emerson's Essays on Thoreau*. North Clarendon, VT: Everyman.

Ulansey, D. (2008). *The current mass extinction*. Online resource database. http://www.well.com/user/davidu/extinction.html. Accessed 9 July 2008.

University of Iowa Center for Human Rights. (2007). #19: Global warming. *Iowa Review*, 37 (2). http://international.uiowa.edu/centers/human-rights/projects/human-rights-index/19-2007.asp. Accessed 18 September 2008.

Warrick, J. (1998). Mass extinction underway, majority of biologists say. In *The Washington Post*, April 21st.

Williams, B. (1981). Persons, character, and morality. In B. Williams (Ed.), *Moral luck*. Cambridge: Cambridge University Press.

Wilson, E. O. (1999). *The diversity of life*. New York: W.W. Norton.

J Agric Environ Ethics (2010) 23:85–108
DOI 10.1007/s10806-009-9187-0

ARTICLES

The Virtue of Simplicity

Joshua Colt Gambrel · Philip Cafaro

Accepted: 27 May 2009 / Published online: 17 June 2009

Abstract In this paper we explore material simplicity, defined as the virtue disposing us to act appropriately within the sphere of our consumer decisions. Simplicity is a conscientious and restrained attitude toward material goods that typically includes (1) decreased consumption and (2) a more conscious consumption; hence (3) greater deliberation regarding our consumer decisions; (4) a more focused life in general; and (5) a greater and more nuanced appreciation for other things besides material goods, and also for (6) material goods themselves. It is to be distinguished from simple-mindedness, a return to nature, or poverty. Simplicity overlaps with traditional virtues such as temperance, frugality, and wisdom, and sustains and enables traditional virtues such as justice and generosity. Simplicity is a virtue because it furthers human flourishing, both individual and social, and sustains nature's ecological flourishing. For analytic purposes, we consider six areas in which simplicity can make important contributions: (1) basic individual flourishing, (2) basic societal flourishing, (3) individual freedom or autonomy, (4) the acquisition of knowledge, (5) living meaningfully, and (6) preserving and protecting nonhuman beings. The proven failure of materialism to secure subjective happiness or objective flourishing argues for the practice of voluntary simplicity and for the radical reform of modern consumer societies.

Keywords Simplicity · Consumption · Temperance · Virtue

J. C. Gambrel (✉)
Blacksburg, VA, USA
e-mail: joshgambrel@gmail.com

P. Cafaro
Philosophy Department, Colorado State University, Fort Collins, CO 80523, USA
e-mail: philip.cafaro@colostate.edu

Introduction

A young folklorist, collecting songs and stories in a remote corner of the southern Appalachians, kept hearing that an old mountain farmer was the keeper of many traditional tales and ballads. Intrigued, he made his way high up into the hills to the man's simple home. After their introduction, the farmer began to sing ballad after ballad and tell tale after tale. From time to time, neighbors, particularly children, would drop by to enjoy his stories.

While the old farmer was singing one of his last songs for the evening, the folklorist noticed a gold rock resting on the front porch where they were sitting. He couldn't help but stare; it was the biggest piece of gold he had ever seen! The farmer noticed this, and when he finished his last song he told the folklorist that he was more than welcome to the shiny rock he had placed on the edge of his porch. The folklorist couldn't believe his ears! After confirming that the farmer was serious, the young folklorist took the gold rock down the mountain to the town where he was staying.

All that evening and through most of the night, he thought about what he would do with the gold when he returned home, and about how crazy that old man was to have given it away. But at dawn, he gathered up the gold rock and headed back up the mountain to the old farmer's house. The farmer, already awake, was surprised when the folklorist handed the gold back to him. "What? I thought you wanted this rock," he said. The folklorist looked at him intently and answered, "Not anymore. Instead, I want to know what you know that made it so easy to give away."[1]

The wise old farmer knows what is important in life and how to live well. To him, sharing the old songs and stories and the companionship that arises from doing so are more significant and life-enriching than gold. His character expresses two key aspects of simplicity: the knowledge of what is truly enjoyable, fulfilling, and meaningful in life; and the wisdom to know when enough is enough. We contend that these two nuggets of wisdom are intimately related, that achieving a good human life is furthered by material simplicity, that because of this, simplicity is indeed a virtue, and that particularly in our time we neglect this virtue at our peril.

Virtue

We Claim Simplicity is a Virtue[2]—But What is Virtue?

"Virtue" is the generic term commonly used for any character trait people wish to commend. In both common speech and philosophical discourse, "the virtues" refer to those qualities whose possession we believe makes a person a *good* person and

[1] This is a regional adaptation of an older Indian story.

[2] We do not, however, claim the primacy of virtue ethics over other general theoretical approaches to ethics, and nothing that follows depends upon this primacy. Any ethical theory will have some conception of human excellence, or virtue; any plausible ethical theory, we believe, should recognize simplicity (in our sense) as a virtue.

able to succeed in characteristic or important human endeavors, and which also help others to do so.

So the virtues do not float free, admired for their intrinsic brilliance. One's particular table of virtues is closely tied to one's particular conception of what makes for a happy, successful, meaningful, or otherwise worthwhile human life—and to what else, beyond their own well-being, we think people should value. The plausibility of any account of virtue depends first on the plausibility or attractiveness of its ideal of a good human life, second on how convincingly it ties possession of certain character traits to achieving that ideal.

There are numerous conceptions of virtue and human well-being. Let us stipulate ours, for the purposes of this paper. Following Martha Nussbaum (1993) and Aristotle, we initially define the virtues as those traits that help promote human flourishing, individually and collectively. On this view, we value both moral and intellectual excellence, since both help constitute and secure human flourishing. On this view, cultivating the virtues is in our self-interest, but not selfish, since doing so helps us to contribute to the flourishing of our families, neighbors, communities, professions, and indeed the whole complex social web that sustains us and that we in turn should work to sustain.

Following Louke van Wensveen (2000) and Aldo Leopold, however, we amend the tradition and define the virtues as those traits that promote human *and nonhuman* flourishing, individually and collectively. We do this because human beings depend on the basic ecological services provided by a flourishing natural world; because we believe human life is enriched by appreciating nature; and because we value nonhuman flourishing as an important end-in-itself. *We* do not flourish in an ecological vacuum. *We* are not all-important.

It is necessary to distinguish individual, social and ecological flourishing for analytic purposes. But in our opinion, all three are so supremely valuable and so intimately related that no human character trait that undermines any one of these counts as a genuine virtue. In the same way, no human institution that undermines any one of these three kinds of flourishing counts as a truly just or fully intelligent institution. We will not defend these assertions here. The joy to be found in listening to the dawn chorus or watching the spring warbler migration, and the dire predictions of the *Millennium Ecosystem Assessment* (2005) and the Intergovernmental Panel on Climate Change's (2007) *Fourth Assessment Report*, if we continue with "business as usual," should be sufficient evidence that humanity does not flourish in an ecological vacuum.

Obviously, amending Nussbaum and Aristotle as Wensveen and Leopold suggest complicates our lives and our moral decision-making. It is easier (and hard enough) to make principled environmental decisions from an anthropocentric perspective. Nevertheless, this amendment is necessary, for the reasons just given. Our flourishing is tied to nature's flourishing; our flourishing is not all-important.

Following Rosalind Hursthouse (who follows Charles Darwin), in this paper, we accept human beings as one more natural kind that has evolved on earth. On this naturalistic view, people are essentially rational social animals, and virtue judgments follow the same logic as ethological judgments about characteristics or traits in nonhuman animals that contribute to their health, normal functioning, or superior fitness. On this view, the virtues, at a minimum, are those human traits that favor (1) individual survival, (2) the continuance of the species, (3) pleasure and freedom from pain, and (4) the good functioning of human communities and societies (Hursthouse 1999, p. 202).

Following Ronald Sandler, however, we amend Hursthouse to provide a fuller account of *human* flourishing and to recognize that human beings, as *rational* animals, may appreciate and work to sustain worthwhile ends that are not directly tied to our own flourishing. So in addition to criteria (1)–(4), we assert that the virtues are those character traits that contribute to (5) human autonomy, (6) the acquisition of knowledge, (7) a meaningful life, and (8) the preservation or promotion of other valuable ends beyond human flourishing (Sandler 2007, p. 28). Such an amended vision of human flourishing and human purposes is more in line with our common moral intuitions. It also coheres better with a naturalistic ethical framework, inasmuch as it presents a more accurate view of what human beings are[3] and of our place in the universe.[4]

Obviously, amending Hursthouse as Sandler suggests complicates our judgments about human flourishing. Whether or not we've successfully met criterion (1) is easier to judge than criterion (7); what to do when criteria (3) and (6) conflict is an open question ("when ignorance is bliss, 'tis folly to be wise," runs the proverb). Nevertheless, this amendment is necessary, to do full justice to human nature and human potential.[5] We are speaking not merely of human survival or continuance, or of that and hedonistic success, but of human flourishing: the *full* development of *all* our essential (or perhaps, possible) capabilities.

In what follows, we argue that simplicity is a virtue, because its cultivation is necessary if we hope to achieve human flourishing, both individually and collectively, in the broad sense outlined above; and to sustain nature's ecological flourishing, also understood broadly. This seems to us the proper framework in which to consider the status and specify the details of any virtue. However, we recognize that conceptions of flourishing and appreciation of nature vary widely, and we believe a strong case can be made for simplicity as a virtue even for those holding quite different conceptions of human flourishing or nature's value.

[3] For example, we can flourish in more complex ways than the other natural beings we know, so that it makes sense to talk about our living autonomously or meaningfully.

[4] Science has led us to understand our relative cosmic unimportance, appreciate the complexity of other life forms here on earth, and recognize our literal kinship with them. It thus helps make the case for the value and importance of nonhuman living beings. There is something pre-Darwinian, even pre-Newtonian, about the anthropocentrism still common among contemporary ethicists (Rolston 1988).

[5] Hursthouse (1999) limits her criteria in an attempt to achieve the level of rational certainty and consensus that botanists and ethologists can achieve talking to one another about plants and nonhuman animals. But this level of certainty and consensus is forever unattainable among those discussing ethics, precisely because people are more complex than other natural beings, and our goals are more open-ended. In other words: virtue ethics does not solve the meta-ethical problems that bedeviled twentieth century ethical philosophy. *This* is not its value. Its value is in allowing us to intelligently discuss human virtue and flourishing. Best to admit this and work to construct a robust account of human flourishing adequate to this (very important) task: not justifying ethics to the skeptic, beyond a reasonable doubt; but rather, providing provisional guidance (*always* partial, dubitable, and revisable) to those already committed to the task of living the best lives possible.

For example, those who would deny that appreciating and promoting nature's flourishing is a necessary criterion of human virtue (rejecting Wensveen's amendment of Nussbaum) must still acknowledge material simplicity's role in furthering the development of a full range of human capabilities. You may be uninterested in species loss and protected by your society's wealth from the most serious dangers of ecosystem collapse (at least temporarily); still, if you are an American, you must navigate the pitfalls of a high consumption society. Personal bankruptcy, unhealthy obesity, and a focus on possessions rather than relationships and achievements, all remain dangerous pitfalls to your flourishing. Again: those who define human flourishing fairly narrowly (rejecting Sandler's amendment of Hursthouse) may perhaps ignore questions of autonomy or meaning, as they focus on the bare essentials of human flourishing. Still, material overconsumption can undermine our hedonistic well-being and the good functioning of society (Hursthouse's criteria 3 and 4) in fairly obvious ways. Health can be undermined by overeating; savings accounts can be emptied out by overspending; banking systems, apparently, can be overwhelmed by greed and its attendant overreaching.

Material simplicity points us toward a nobler conception of human flourishing and facilitates its accomplishment. But it also improves our chances of achieving more modest or conventional forms of well-being. It is a virtue for eco-saints and moral strivers, but also for just plain folks.[6] This strengthens its claim to virtue-hood. In what follows, we ask our readers to consider material simplicity's potential contributions to happiness, flourishing, or human well-being, however *you* understand them, and over the full range of what *you* consider to be morally acceptable kinds of good human lives; also (if this matters to you), simplicity's potential contribution to the continued survival and flourishing of the five to ten million other species with whom we share the earth. Your assessment of these contributions should control your answers as to whether or not simplicity is a virtue, its relative importance in our lives, and its further detailed specification.

Simplicity

In her seminal article "Non-Relative Virtues: an Aristotelian Approach," Martha Nussbaum provides a useful way to define and distinguish the virtues. "Isolate a sphere of human experience that figures in more or less any human life, and in which more or less any human being will have to make *some* choices rather than others, and act in *some* way rather than some other," she suggests. Crucially, to require the specification of a virtue or a range of virtuous behavior in this area, these choices must be important to people's happiness or well-being. "The 'thin account' of each virtue is that it is whatever being stably disposed to act appropriately in that sphere consists in" (Nussbaum 1993, p. 245). The "full or 'thick' description" of the virtue examines the opportunities and pitfalls that await in the realm of

[6] Compare courage, a virtue for the soldier facing death on the battlefield and for the nervous applicant looking to make a good impression in a job interview.

experience under analysis. It specifies the characteristic thought processes, habituation, and emotional development, ways of looking at the world, and other aspects of human character and training that help us choose well in that particular sphere. And it ties choosing well to *living* well and flourishing as a human being.

Following Nussbaum's schema, we define simplicity as the virtue disposing us to act appropriately within the sphere of our consumer decisions, from food and drink to stereo and housing purchases to cars and airplane travel. As we understand it, simplicity is a conscientious and restrained attitude toward material goods. It typically includes (1) decreased consumption and (2) a more conscious consumption; hence (3) greater deliberation regarding our consumer decisions; (4) a more focused life in general; and (5) a greater and more nuanced appreciation for other things besides material goods, and also for (6) material goods themselves.

As Aristotle noted long ago, people may be much more likely to err in one direction rather than another in particular spheres of human choice, either due to human nature or to the pathologies of their particular societies (*Nicomachean Ethics*, book II, Chap. 8). In Athens in Aristotle's day, men were apparently more likely to err on the side of irascibility than "inirascibility"; hence Aristotle named the virtue with regard to anger "mildness" (book IV, Chap. 5). In wealthy western democracies today, we contend, people are more likely to err on the side of overconsumption than underconsumption. Hence the term "simplicity" is arguably a good one for this virtue (as long as we remember that underconsumption can also be a problem).

Simplicity overlaps with such traditional virtues as temperance (moderation in food and drink), frugality (the responsible and restrained use of wealth), prudence, and self-control.[7] In what follows, we make no attempt to systematically relate our treatment of simplicity, focused on overall material consumption, to traditional discussions of these virtues, which often ranged more narrowly or broadly than ours; and which took place in comparatively poor societies whose economies of scarcity have been replaced by economies of abundance throughout the modern industrialized world. We note, however, that within the tradition, philosophers once routinely claimed that temperance, frugality, and simplicity were keys to living justly and wisely.[8] They were right. We see it as a glaring weakness of contemporary discussions of justice and wisdom that they rarely make this connection.[9]

It is often helpful to consider simplicity as a virtuous mean between vicious extremes. However, like other complex virtues, simplicity appears to be a mean

[7] These virtues, in turn, overlap in the tradition. Note for example that *sophrosune*, the virtue explored in Plato's *Charmides*, may be translated into English as either "temperance" or "soundmindedness," the latter a close kin to prudence.

[8] For a start, see Plato, *Apology* and *Republic*.

[9] Valerie Tiberius's (2008) otherwise excellent book on wisdom makes no connection between material simplicity and her subject. You may find "Terrell, Huntington," in the index to John Rawl's *A Theory of Justice*, but not "temperance." Greater attention to the real world impediments to wisdom and justice would lead philosophers to pay more attention to material simplicity.

along several axes. Some of its associated vices have obvious names, others do not, perhaps owing to their rarity:

Vice	Virtue	Vice
	Simplicity	
Underconsumption (poverty?)		Overconsumption (gluttony)
Unthinking consumption (carelessness)		Overthinking consumption (obsession)
None; or crude consumption (asceticism, "monkish virtue")		Luxurious consumption
Inefficient or pointless consumption (wastefulness)		Hyper-efficient consumption (Penny-pinching)
Immoral consumption (callous, disproportional)		None; or, moral finickiness ("moral foppery")

Obviously, there is more than one way to go wrong in our stance toward consumption!

An Example

Treating simplicity as a virtue presupposes that through reflection, we can discover our deeper, more significant needs and goals; recognize some goals as ignoble, foolish, or trivial, and replace them with better ones; and pursue our goals more efficiently, with less waste and harm to others. By way of illustration, consider some steps a person might take to practice voluntary simplicity in relation to food consumption, as these relate to the six aspects of simplicity noted above.

Americans consume on average 25% more calories than necessary, on a conservative estimate (Putnam et al. 2002). Today, three out of five Americans are overweight and one out of five is obese. This excessive consumption of food harms our health and quality of life (US Department of Health and Human Services 2001). Food overconsumption also causes direct and indirect environmental degradation, through habitat loss and increased pollution from agricultural fertilizers and pesticides (Cafaro et al. 2006). Twenty percent of American greenhouse gas emissions come from growing and transporting our food (Pollan 2007). So here simplicity clearly demands decreased consumption (aspect 1).

However, whether we are talking about personal health, healthy communities, or healthy land, consuming less is not enough. We also need to consume *differently*.[10] We may buy and prepare more healthy foods for ourselves; buy organic foods and local foods and eat less meat, all of which decrease environmental harms; purchase more food directly from farmers at farmers' markets or as part of community-supported agriculture co-ops, to support small farmers and keep local agriculture

[10] This point is emphasized by Crocker (1998), who uses it to de-emphasize the importance of consuming less—mistakenly, in our opinion.

vibrant. Such changes demand attention: a more conscious consumption, involving greater deliberation about our consumer decisions (aspects 2 and 3).

Many food simplifiers combine more conscious consumption with greater participation in food *production*: gardening, raising chickens, keeping bees, or joining CSA cooperatives. Research shows that food produced in these ways is more environmentally sustainable and often more nutritious than conventionally grown food (Felice 2007). These activities are also often enjoyable and interesting, and connect people to their neighbors and to the earth. Similarly, taking time to prepare our own food and eating meals together offer important opportunities to connect to loved ones. Consciously taking such steps leads to a more focused life (aspect 4). It can further develop gratitude toward the many other species that sustain us; tune us into nature's rhythms and details; and enrich our relationships with other people. In these ways, food simplifiers explore and sustain a wide range of nonmaterial goods (aspect 5) and come to better understand and appreciate the material realities of food production and consumption (aspect 6).

The example should begin to suggest how simplicity can contribute to human and nonhuman flourishing in important ways. It also illustrates several important points about simplicity as a virtue.

First, living simply is not necessarily simple. It requires deliberation: thinking through our choices and acting on our best judgment, rather than following the herd, or the blandishments of advertisers, or doing what we have always done, or what comes easiest.[11] Thinking about our food consumption and improving it typically involves research and planning, and some of what we learn about how our food is grown will probably be discouraging or disgusting. Still, it is better to know the ugly facts and act in full consciousness of what we are a part of, rather than in ignorance. Simplicity is better than simple-mindedness, the default setting of the American food consumer (an ignorance that the food industry spends many millions of dollars a year cultivating).

Second, though, simplicity *is* often simple; or rather, it often involves working our way back to simpler, less convoluted ways of doing things. When I plant and tend a garden, ride a bicycle and fix it myself, or sing songs with my children on family outings, these are relatively simple ways of satisfying some of my food, transportation, and entertainment needs. They reverse our modern tendency to an ever more elaborate division of labor; "a principle," as Henry Thoreau once said, "which should never be followed but with circumspection" (Thoreau 1971). The simplicity of such activities makes them less likely to stray from their goals and more likely to involve thoughtful activity rather than passive consumption. Their simplicity, the fact they we can "see all the way through them," may make them particularly appropriate vessels for finding meaning, or expressing happiness and gratitude.

Third, simplicity is not a call to "return to nature" in any romantic or primitivist sense. Old ways can be wasteful, or harmful; new ways can be an improvement. Similarly, simplicity is not opposed to technology, or to new technologies. It just

[11] For this reason, Comte-Sponville (2001) errs in describing simplicity as "an antidote to introspection" and imaging this virtue as a sort of primitive integrity.

asks that we consciously develop and appropriately incorporate technologies into our lives with reference to our real purposes and to their full effects on the world around us. Hydroponics has a role to play, along with sharing heirloom tomato seeds with the neighbors.

Fourth, simplicity is not poverty. Poverty is a state defined by lack, where people find it difficult to obtain the means to satisfy even the essential human needs—food, water, shelter, basic physical safety—let alone higher needs for self-actualization or creative personal development. Poverty means living in deprivation, against one's will. Simplicity is consciously and freely chosen.[12] It provides greater opportunities than conventional materialism to achieve human flourishing, while poverty limits those opportunities.

Fifth, simplicity is a process, not an endpoint. Although we are *arguing* here for simplicity, we have not forgotten Aristotle's reminders that habituation is more important than arguments in developing virtue; and that virtue demands *phronesis*, practical wisdom, applied to the details of life. Anyone who has tried to cultivate simplicity in their own life knows that Aristotle was right. Creating a character, a personal infrastructure, and daily habits that regularly result in less consumption and less dumb consumption, are difficult and ongoing affairs. Hence it is a mistake to look for particular markers that indicate the presence of this virtue (although a Hummer in the driveway pretty clearly indicates its absence. And it is not a mistake, but good practice, to set down markers for ourselves and strive to achieve them).

Sixth, and related to the previous point, simplicity is not uniformity. There are as many ways to cultivate simplicity in our food or other consumption decisions as there are ways to complexify them. Different people will focus on different aspects of these problems, and our solutions should play to our individual interests and strengths (maybe you'd rather brew beer than raise tomatoes; perhaps you're the cook, not the gardener, in the family). Hence lives and lifestyles will legitimately differ. Simplicity need not limit diversity.

Seventh, simplicity, like all the virtues, needs to be cultivated by individuals and families, but also encouraged and sometimes mandated by society, if we hope to secure human and nonhuman flourishing. The very term *voluntary* simplicity emphasizes voluntarism, while most of the literature on material simplicity focuses on individual and small-group action. But this is arguably a failure of this literature (Claxton 1994). Jerome Segal (1999) argues convincingly that creating less materialistic societies will demand fundamental political changes. Discussing the United States, Segal emphasizes changes in economic policy that would help safeguard basic physical and economic security, and thus make it easier for individuals to freely choose less materialistic paths. Because we often "consume because others consume" (Lichtenberg 1998) and because "what counts as necessary [consumption] in a given society" depends in part on "what the poorest members of society require for credible social standing (Schudson 1998), enacting simplicity has an important political component.

[12] However, we want to leave room for *communities* to choose simplicity, not just individuals. We do not agree with a purely voluntaristic view of the cultivation of virtue. The primary purpose of law is to force citizens to act more virtuously than they otherwise would.

Eighth, we are not arguing that we should eliminate or minimize material goals within our lives. Human beings have essential needs that must be met, so we must devote a certain amount of our time and attention to meeting them. Simplicity actually represents a better way of meeting these needs than high levels or unthinking kinds of consumption. In the case of food, "better" means in ways that are healthier, more morally justifiable, and more interesting, enjoyable, and meaningful than business as usual. Simplicity can help us really understand our material needs and meet them efficiently, morally, and even joyfully.

Ninth (and at the risk of sounding grandiose), our short discussion of food simplicity suggests that material simplicity does indeed further justice and wisdom, as philosophers have long maintained. Modern industrial agriculture is callous toward farmers and farm communities, and grossly unjust toward its animal "production units." These injustices are sustained in large part by the ignorance of consumers. Voluntary food simplicity can help reverse this process, as we learn about food and act on what we have learned, try to appreciate the processes involved in feeding us, and honor the various participants in those processes (Berry 1990). To the extent we use resources, take life, or cause pain when we raise or eat food, simplicity enables us to do so consciously and honestly. This opens up a space within which we *may* act justly and wisely. Note that simplicity does not *guarantee* justice and wisdom in this important area of our lives (or in others); it makes them possible.[13] Casual participants in the industrial food status quo, however, cannot act justly or wisely in their food consumption decisions. Those options are not on the menu.

True, few people in our society would look to a person's eating habits to discern whether they are just or wise. But we believe this is a measure of how distracted we have become and how confused about what is really important in life.

Again, others might say that what are really needed are better rules for how farmers should be compensated, food animals treated, and so forth. Then we could follow the rules and eat whatever we wanted, with a clean conscience. We certainly agree on the need for better rules: they are essential to furthering material simplicity politically, and thus helping create more virtuous and just societies. But rules will only get us so far. The world is an unjust place and seems likely to remain so for the foreseeable future; hence we cannot completely rely on "the rules" to tell us how to behave.[14] Further, the idea of purely economic spheres of life, where we can choose freely—that is, without the need to consult anything but our own desires and whims, perhaps restrained by a few basic moral rules—is deeply flawed; part of the economistic view of life that has given us modern industrial agriculture in the first place. Setting up such "duty free zones" blinds us to both responsibilities and opportunities. We think we are increasing our options and our freedom of action; instead, we find we have lost the ability to distinguish right from wrong, or quantity

[13] As one reviewer put it: "simplicity is more like a clearing of the stage. It judges conspicuous consumption" or gluttony "to be not worthwhile, and so eliminates some practices. But it does not itself put a worthier goal in place of consumption." We might say that simplicity is an *enabling* virtue.

[14] Not to mention that rules can neither enact nor apply themselves, facts discussed extensively in the recent philosophical literature.

from quality. But acting on such distinctions, in all areas of our lives, is the very definition of wisdom.

To conclude this section, though, let us descend a bit from the heights. Most people don't aspire to wisdom, and most of us are content to exhibit about the same degree of justice as our neighbors. Most of us concentrate on our own subjective happiness most of the time. Even so, even from this low perspective, we assert the proven failure of materialism. While Americans' average incomes increased from $9,000 in 1957 to $20,000 by the end of the twentieth century (in constant dollars), the percentage describing ourselves as "happy" or "very happy" in national polls declined steeply during this time. Maybe it's time to give material simplicity a try.

The Case for Simplicity as a Virtue

Simplicity is a virtue, we claim, because it furthers human and nonhuman flourishing. For analytic purposes, we next consider six areas in which simplicity can make important contributions: (1) basic individual flourishing, (2) basic societal flourishing, (3) individual freedom/autonomy, (4) the acquisition of knowledge, (5) living meaningfully, and (6) the flourishing of nonhuman beings. We'll address each of these areas in turn. While we primarily consider simplicity in relation to the consumption options of middle-class Americans, we note that high consumption lifestyles are becoming an actual or aspirational norm across much of the world. Furthermore, the issues we discuss, if not their details, remain important across widely diverse societies and social classes. Readers may decide for themselves how much of what follows applies to their own consumption decisions, or their society's, and the right way to apply it.

Basic Human Flourishing

Simplifying our lives and lessening consumption have the potential to significantly improve human health. As already mentioned, Americans on average eat too much; we also often eat nutritionally poor food. Four of the top ten causes of death are directly related to overweight and obesity, including the top three: heart attack, stroke, and cancer (Cafaro et al. 2006, p. 543). In addition to increasing morbidity and mortality, overweight/obesity have further effects that are harmful to happiness, from reducing people's sex drives to lowering their self-esteem. Emerging evidence also suggests that poor nutrition often plays a role in causing depression (Jacka 2007). Nutritionally poor diets influence relevant aspects of our endocrine development, resulting in a weakened stress response system, which in turn makes us more vulnerable to many ailments, such as insomnia, hypertension, asthma, anxiety, and alcoholism (Mokdad et al. 2001). The moderate consumption of essential nutrients modifies the synthesis, storage, release, and action of neurotransmitters in the central nervous system, which makes us less susceptible to psychological disorders resulting from stress (ibid.).

For all these reasons, conscious and temperate food consumption is a key to health. And health, in turn, is the basic grounding for human happiness and flourishing.

Financial over-commitments are another leading cause of unhappiness in modern America. Part of material simplicity is old-fashioned frugality: not buying things unless we are sure we can afford them and there is some rational justification for doing so (Nash 1998). Holding our consumer purchases to this standard will result in fewer purchases. This in turn will mean less worry about their maintenance and care—not to mention fewer payments! The money freed up can go toward personal savings, or providing increased financial security. Alternatively, people with some flexibility can reduce their work hours, freeing up time to spend with their families or in other enjoyable activities (Segal 1998). Either way, we would often increase our happiness by not wasting money on unnecessary stuff (or on unnecessarily expensive stuff). As we spend less time working to pay off debts accumulated from habitual material upgrades and wanton expenditures, we should be able to find more time to cultivate our own passions and visions, and to engage in activities that are truly enjoyable.

As Americans have worked longer hours in recent decades, we have also become more depressed (Cross-National Collaborative Group 1992). Americans work on average 150 more hours per year than we did 50 years ago, and we are ten times more likely to suffer from depression (De Graaf et al. 2002). By passing on the new car or expensive vacation, we could work less. Not only would this be good in itself, it would free up time to play with our children, talk to our spouses, or volunteer in our communities. These sorts of activities strengthen our personal relationships, and psychological studies strongly confirm that positive personal relationships—unlike material wealth—are key to human happiness (Lane 1998; House 1986).

Getting these "basics" right is important—but not all important. Positive psychologists, going back as far as Maslow (1999), have argued convincingly that full human flourishing is experienced in self-actualization and personal growth, after all our basic needs have been met. Higher states of human development involve experiences of transcendence and the appreciation of beauty, artistic creation, the pursuit of knowledge, and spiritual transformation. In these higher states we realize our full potential as human beings and discover the potential for creativity even in mundane activities (Peterson and Seligman 2004).

Simplicity makes possible such self-actualized activities and states, in part because it frees up time to pursue them, in part because it provides a more salubrious and effective manner of fulfilling our more basic needs. When we cultivate simplicity, we are more likely to feel assured that our physiological and security needs, our affiliation and affection needs, our self-respect and recognition needs, have all been adequately satisfied. This allows us to move on to our higher-order growth needs. A materialistic perspective, on the other hand, tends to leave us unsatisfied, *no matter how much material success we achieve* (Kasser et al. 2004). Hence we remain insecure, and fail to move beyond the level of basic needs.

Basic Societal Flourishing

Cultivating simplicity also promotes societal flourishing. First, the perpetually dissatisfied attitude fostered by materialism contaminates and undermines our human relationships. The higher married couples score on a standard psychological "materialism" scale, the more likely they are to get a divorce. Materialistic couples also report lower-quality relationships with their partners. According to one researcher: "when people place a strong emphasis on consuming and buying… they may also become more likely to treat people like things" (Kasser 2002). Half of all American marriages now end in divorce; importantly, family counselors claim that 90% of these separations involve financial stress (Bennett 1993). This financial stress could be lessened by reducing our material wants. There is no better indicator of a healthy society than the state of the social unions that comprise it.

Materialism also harms children. Research shows that materialistic parents tend to be less nurturing and supportive than less materialistic parents, leading their children to idolize materialistic social icons and develop feelings of insecurity (Kasser 2002, pp. 30–32). Materialistic parents also tend to have materialistic children, as "experiences that undermine the satisfaction of psychological needs can cause individuals to orient toward materialism as one type of compensatory strategy intended to countermand the distressing effects of feelings of insecurity" (ibid.). But value orientations that stress self-development, creativity, or building social connections, provide more effective sources for children's well-being. David Goleman's (1995) research shows that parents who are less materialistic tend to be happier and have happier, more socially adjusted children.

Simplicity may also give rise to a more developed sense of citizenship and communal responsibility. Instead of pursuing material goods, we may spend our time volunteering and strengthening community institutions.[15] Americans today are often "time poor" (Schor 1992). By freeing up more time to spend with loved ones, neighbors, and community members, simplicity can help us develop the social unions that enrich our lives. By fostering contentment with our status and possessions and reducing levels of dissatisfaction, simplicity can help minimize social tension and build up social capital.

Finally, we need to remember that healthy societies depend on adequate ecosystem services, which are threatened by present consumption levels (Millennium Ecosystem Assessment 2005; IPCC 2007). Our current overconsumption cannot continue indefinitely and needs to end soon—even if only in our own self-interest. Of course, reducing humanity's ecological footprint cannot be just a matter of individual voluntarism; whole nations need to consume less, a reality we are far from facing at the present time. Voluntary simplifiers perhaps provide a greater social service not through the direct effects of their lower consumption, but by showing their fellow citizens that consuming less need not be a horrible imposition, but may actually improve their lives.

[15] Note that material simplicity doesn't *guarantee* any of this. But it does make it more likely. In our observations, fathers who coach their kids' soccer teams, mothers who chaperone school field trips, community activists who show up for city council meetings, all tend to be less materialistic than average.

Freedom and Autonomy

Freedom is a basic human good and a fundamental human desire. Material possessions sometimes increase and sometimes decrease our freedom; it requires wisdom to decide which, in particular instances. Henry Thoreau claimed that "a man is rich in proportion to the number of things he can afford to leave alone"; he is free, that is, both from the craving for those things and from having to earn the money to buy them. But even Thoreau appreciated being able to take the train from Concord to Boston, rather than walking back and forth every time he had an errand there. He appreciated the freedom a well-built canoe provided in the Maine backcountry.

Full human freedom includes the ability to see and set limits to our pursuit of material goods. We do not deny the value of some stuff for increasing freedom. But we insist that consciously deciding the role of possessions in our lives is one important key to freedom. Whether or not "the market" brings you your goods, letting the market *society* decide what goods you need shows a lack of freedom and can lead to fear, anxiety, and unhappiness (Kanner and Soule 2004). Unfortunately, in America, ubiquitous advertising preaches dissatisfaction and undermines our ability to think intelligently about these matters. One hundred and sixty-three *billion* dollars in advertising was directed at the American public in 2006—almost half the total world advertising budget!—in order to keep us consuming at high levels (World Advertising Research Center 2007). We are overwhelmed by innumerable consumer choices, but as Barry Schwartz's (2004) research has shown, as our material choices increase beyond a certain point, people actually become less happy, less optimistic, and more depressed. By preaching dissatisfaction, advertising leads us to believe that in order to be satisfied we must increase consumption—which ensures less satisfaction. In real ways, the treadmill of over-consumption restricts our freedom and autonomy as individuals.

Take the important example of the automobile, a powerful symbol of freedom for many people, particularly Americans.[16] Today most Americans simply expect that every adult should own their own car and drive it whenever and wherever they want; anything less is seen as seriously cramping our freedom. In reality, given common financial constraints, not owning a car can free up vital resources for students and poorer workers—provided they can get where they need to get without one. Commuting by bicycle or train can be more relaxing and enjoyable, even for wealthier commuters. Millions of people have these options but are unaware of them, so automatic have their transportation decisions become. This cluelessness limits their freedom. Similarly, the development of auto-centric urban areas and our society's failure to provide adequate mass transportation may force many people to buy, drive, and maintain cars who would have preferred not to. It may take away their freedom *not* to drive, *not* to make car payments, *not* to subsidize environmental destruction by the oil companies.

All this is not to say that every American should junk his or her car. We only suggest that in some cases, limiting our automobile use, either individually or

[16] For a more detailed discussion of consumption and cars, see Cafaro (1998).

collectively, may make Americans more free—but that for the most part, we are unable to think clearly about this, because our society tends to confuse us concerning the connections between freedom and consumption. At a minimum, we should try to recognize all the consequences of our transportation and other consumption choices. Acting on the basis of such recognition, personally and politically, is an important part of freedom.

Knowledge

Simplicity also can aid in the acquisition of two important kinds of knowledge. First, simplicity assists us in becoming more self-aware. Simplifying our lives requires us to look within ourselves to distinguish our true needs and most important goals from what is inessential or positively harmful in our lives. It may force us to confront some of our less charming weaknesses and obsessions and some of our more important moral failings. Attempting to practice simplicity can force us to find the proper balance between demanding more from ourselves and showing charity toward our own failings.

Know thyself! It is an essential human task and will help you live a better life, however, you define that. People with greater self-knowledge are better positioned to achieve happier and more meaningful lives and develop the full spectrum of human potentials: physical, emotional, mental, and spiritual. Self-knowledge facilitates the self-control necessary to combat materialism and live more focused lives. All this increases our ability to flourish.

Second, simplicity facilitates local, place-based, ecological knowledge. Such knowledge is good in itself, as Aldo Leopold (1949) illustrates memorably in the first two-thirds of *A Sand County Almanac*; it is also the backbone for much environmental activism. Make this simple experiment. Select a stretch of road that you have often traveled by car, and bicycle along it instead. Next, take even more time and walk it. One of this article's authors once did this along a 10-mile stretch of rural highway in coastal Oregon. Having roared down it every day for a whole summer to and from work, the next summer he traded in his motorcycle for a bicycle, commuting along the same route. There were a thousand and one things that he only noticed once he began to bicycle. People, houses, trees; smells, sounds; the pitch of the terrain, the different water levels in the river at high and low tides. Not all that he noticed was pretty. There were trash dumps and junked cars. But traveling slower, he noticed new construction and shoddy logging practices along the way that made possible informed participation in local zoning hearings and town debates.

This is just an example, but we find it suggestive. Because he no longer owned a motorcycle, your author did not have the choice of driving to work, and was forced to bicycle on days when he often would not have. Here less choice equaled more experience, and less convenience equaled greater knowledge of the local landscape. Modern microeconomics unequivocally says that more choice is better: give an individual a greater number of choices and more money to spend on them, and he can more fully satisfy his preferences (Heyne 1994, p. 178). Modern advertising sends the same message: look at the many new, improved cars, computers, or shaving creams to choose from! But this is unconvincing. For we can choose wrongly, and in ways that lead to less pleasure, less health, and less knowledge of the world around us. One of the ways we can choose wisely is in deciding to limit our options and leaving them limited. Choosing correctly also involves limiting the amount of time we spend choosing, thus leaving our minds free to deal with more important things (Schmidtz 2008). We believe it involves simplifying our lives so that we maximize knowledge and appreciation of the world, rather than consumption and wealth.

Meaning

Cultivating simplicity can also enhance meaning in our lives by connecting us in rich ways to our inner hopes, visions, and potential. William Frankena thinks finding meaning in life involves "being convinced of the importance of what we care about." Living meaningfully also involves acting in service to goals and ideals that we are convinced are important. Living a life devoted to things and the status they provide is deeply unsatisfying to many people. What people tend to find most meaningful is having good relationships with other people and achieving personal and career goals that, on reflection, make them proud. Materialistic pursuits tend to trivialize our lives, instead of enriching them with meaning. We have already seen that materialism increases the likelihood that we will grow up to be non-nurturing parents, that we will live in fear, and that we will doubt our own self-worth. In order to connect to people and activities that we find meaningful, it helps to get off the materialistic treadmill.

In this regard, we should remember our natural tendency to use material acquisitions to assuage perceived deficiencies. One-third of all shoppers are "shopping therapeutically"; that is, they are shopping to feel better about themselves (De Graaf et al. 2002). Unfortunately, this often leads to further dissatisfaction and discontent (Kasser et al. 2004; Lane 1998). If we keep in mind what human flourishing really requires, we'll remember that money and possessions are merely means, of limited usefulness, and not ends in themselves. Money cannot buy self-esteem, wisdom, or love. It cannot buy the friendship or respect of people whose friendship and respect matters. Before we purchase anything, we should ask ourselves what contribution, if any, the purchase might make toward meeting our highest goals or important projects. We should ask whether our material purchases bring meaning into our lives.

Promoting the Flourishing of Nonhuman Beings

Simplicity promotes human flourishing. Just as clearly, it is necessary to create and preserve space and resources for nonhuman species and ecosystems to flourish (Wensveen 2001). Due primarily to overpopulation and overconsumption, humans are extinguishing animal and plant species a thousand times faster than natural background extinction rates (Millennium Ecosystem Assessment 2005). Along with many scientists and environmentalists, we believe that this loss of biodiversity is the most significant environmental problem facing humanity today. The ability of our global biosphere to sustainably provide for human and non-human flourishing is stressed by continued population growth and by growing global overconsumption. These two trends are negatively synergistic; if we don't reverse them, we may lose up to one-third of the world's species of plants and animals within 50 years (ibid.).

In the face of this overwhelming extinction threat, most people recognize the need to "do something." Buy and set aside more public lands as wildlife habitat; find alternative technologies to limit air and water pollution. These steps are necessary, but not sufficient. We cannot set aside land if we need it to feed our appetites for food, fuel, or other material goods (Cafaro et al. 2006). Technological improvements can alleviate pollution, but without some limits to human demands, the improvements are swamped and pollution increases (as happened in the US over the past 30 years with carbon emissions, as increased energy efficiency was swamped by greater per capita energy demands and by ever more "capitas"; Camarota 2008).

Limiting consumption *must* be part of the solution to the extinction crisis and the other environmental challenges we face in the 21st century. Yet it is a part of the solution that we are very unwilling to face.

Consider global warming. Stephen Pacala and Robert Socolow's "wedge diagram," below, is a heuristic designed to help us think about the steps needed to address global warming (Fig. 1):

Each "wedge" in the "stabilization triangle" above represents a change in human resource use which, if fully implemented, would keep one billion metric tons of

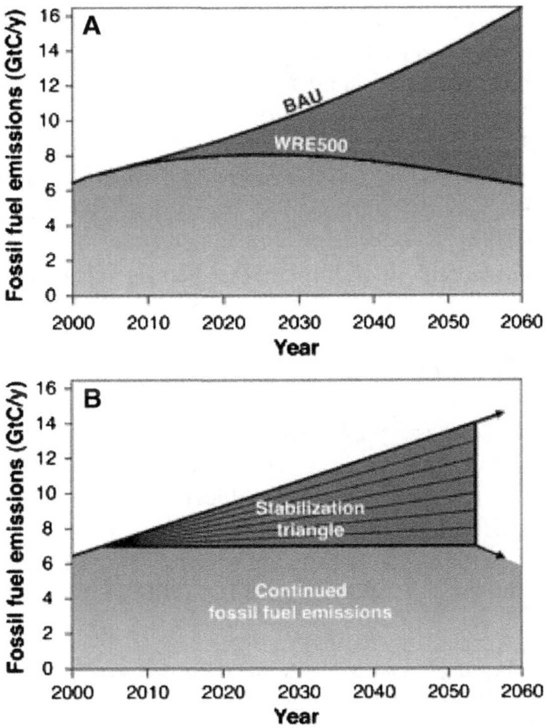

Fig. 1 Source: Pacala and Socolow (2004)

carbon from being pumped into the air 50 years from now. The authors reckon *eight* such wedges must be implemented—not to reduce atmospheric carbon; not to stabilize atmospheric carbon levels—but simply to keep atmospheric carbon from pushing past potentially catastrophic levels during this period.

The key point is that of the sixteen wedges that Pacala and Socolow (2004) propose as most feasible for reducing carbon emissions in a big way, *fifteen* focus on technological changes in energy production (more solar, more nuclear), energy efficiency (increasing average auto fuel efficiency from 30 to 60 mpg), or land use (conservation tillage on agricultural lands); and only *one* focuses on limiting human consumption (cutting the average number of miles driven by automobile drivers in half).[17] One out of 16! Most discussions of global warming are similarly biased toward "technofixes." An influential example was a 2006 proposal for US climate action developed by scientists affiliated with the Natural Resources Defense Council (Lashof and Hawkins 2006). Again, limiting consumption was hardly mentioned; the goal, as always, was to accommodate *more* consumption by *more* people with *less* environmental impact.

Such approaches, in their unwillingness to face the need for people to live more simply, are puerile. More important, *they are unlikely to succeed.* In addition to seeking technological improvements, human societies must find ways to embrace less consumption—or face the consequences in environmental degradation. Environmentalists of all kinds—anthropocentrists and biocentrists, urban environmental justice advocates and wildlands proponents—should all be able to agree on the centrality of reining in human consumption to secure our goals.

Environmental protection is a special case—and a supremely important case—of the more general rule, that individuals and societies that can intelligently moderate their consumption open up a space to preserve and enhance their preeminent values. Preserving the flourishing of nonhuman species should be one of those values. However, even people who do not care about protecting wild nature tend to care about the safety and security of their children and grandchildren. So we conclude this section with the reminder that evidence is rapidly accumulating that our posterity's well-being will depend on keeping human demands on the biosphere within ecological limits (IPCC 2007). That means learning to live more simply.

Objections

In all these ways, we contend, material simplicity helps further human and nonhuman flourishing. For these reasons, simplicity is a virtue. Let us now turn to a few of the most common objections to this position.

One objection runs, "If simplicity provides such wonderful benefits and materialism is so flawed, why are so many people materialistic? There must be something to this approach to life!" We answer that overconsumption and a

[17] Incredibly, *none* of Pacala and Socolow's wedges focus on limiting population growth, despite the critical role it plays in causing global warming (Intergovernmental Panel on Climate Change 2007). For a discussion of the need for environmentalists to address population issues, see Cafaro and Staples (2009).

materialistic value orientation are heavily promoted in America and elsewhere through advertising, popular culture and our education systems (Richins and Dawson 1992; World Advertising Research Center 2007). This onslaught is more than enough to account for prevailing levels of materialism. Research has also shown that when our psychological needs are not being met in healthy ways, we often substitute material acquisition as a coping strategy to counter the distressing effects of insecurity (Kasser 2002; Kottler et al. 2004). America has higher levels of economic insecurity than most other wealthy industrialized democracies; for example, because we fail to guarantee decent health care for all our citizens. We believe this pushes Americans toward greater materialism. In effect, we adopt materialism as a way of compensating for various inadequacies, either personal or social.

In another possible objection, a devoted urbanite might wonder, "Is simplicity antithetical to urban living?" Despite what our food-focused examples might have suggested, we answer: not at all. As we commit ourselves to simplicity, we will discover that we can cultivate this virtue equally well in the city or the country. Many people who live in larger cities utilize public transportation, eliminating the need for a car. They tend to have necessary services located nearer to their homes or apartments, reducing travel and driving times. Larger concentrations of people may increase opportunities to socialize, or to enjoy a greater variety of cultural activities. "Citizens" also have smaller homes and thus less room to store superfluous possessions and less space to heat and cool. Anyone can cultivate simplicity, anywhere, as long as they commit to living a deliberate life. Materialism and overconsumption tend to impoverish us, regardless of where we live.

A third objector might ask, "What if someone isn't interested in cultivating capacities to fulfill her higher needs? Instead, we may imagine, she is happy having her basic needs met in ever more luxurious ways." There is no proof of the irrationality of this approach to living. Still, we would argue with the sybarite that throughout history, simplicity has proven itself more effective than luxury in satisfying people's basic needs.[18] Materialism *tends* to lead toward overconsumption, which results in many of the problems we've already discussed. And even when our basic needs can be met in luxurious ways, simplicity frees up resources that allow other people and other species to flourish. Fulfilling our basic needs in ever more opulent ways is arguably selfish, and hence open to the standard moral arguments against selfishness.

What of a fourth objector, who accepts the Aristotelian/Maslovian position that human beings should pursue higher goals than mere pleasure or subjective happiness, but who sees no problem with living high on the hog while doing so? There are plenty of artists who live for their art, scientists focused on the pursuit of knowledge, writers devoted to their craft, and philosophers seeking enlightenment regarding the big questions—who are happy to eat *foie gras* and drink Dom Perignon champagne when the opportunity presents itself, or jet around the world to

[18] For the classic argument, see Epicurus, *Letter to Menoeceus* and *Principle Doctrines*. An excellent updating of Epicurus from a modern environmentalist perspective is Stephanie Mills' *Epicurean Simplicity* (2003).

take part in symposia on art, science, literature, or philosophy at expensive resorts. Indeed, those at the top of their fields often enjoy such getaways as a perk of success, a necessary break from their work, and an opportunity to network and enjoy themselves, while those with more middling success still often consume at the highest levels they can afford.

We respond that plenty of artists, writers, and even philosophers have fallen prey to their own poor consumption decisions, undermining their health, financial well-being, personal relationships, or careers. Those pursuing particularly ambitious creative goals may have special reasons for practicing material simplicity, in order to husband financial and psychic resources for their efforts. And of course, the standard moral arguments for fair sharing of resources with others still apply to the creative and successful among us. So simplicity remains a virtue, we think, even in these cases. However, we are not committed to a single vision of human flourishing, or to quibbling about particular consumer purchases; nor do we advocate the complete triumph of Apollo over Dionysus. Our main point is that simplicity allows people to meet their material needs in healthy, equitable, and ecologically and financially sustainable ways.

In a fifth objection, some claim that simplicity is only a virtue for middle-class or rich members of wealthy societies. As one reviewer wrote: "It would not go over well to walk around the slums of Mumbai and tell people that they should practice the virtue of simplicity to help them free up time to pursue self-actualized states and activities." Of course, most of our readers are relatively well-to-do members of wealthy societies. So we could simply accept this objection and continue to advocate for material simplicity, with the caveat that we are not speaking about the choices faced by poor people. This would probably be the less presumptuous and more rhetorically effective response.

However, we are not sure it would be the correct one. It goes without saying that poor people face different consumer decisions than rich people. It also seems clear that compared to rich people, poor people will more often be rationally justified, on prudential or moral grounds, in striving to *increase* their consumption. Still, poor people, like rich people, need to make intelligent consumer decisions; in fact, they may have even greater need to choose wisely and with discipline, because they have a smaller margin of error. Consuming less may allow poor people to stretch their limited resources further, to better support their flourishing and the flourishing of their families. Consuming too much, in the absence of the safety nets safeguarding wealthier classes or nations, may lead to disaster. So the virtue we have defined as "simplicity" arguably remains a virtue for poor people and perhaps even for very poor people, although the word may no longer be appropriate (perhaps at some point "discipline" better captures what is needed in the realm of consumer choices; perhaps as people's consumption options narrow, "temperance" and "frugality" say all that needs to be said).

Note that it is no part of our argument to try to justify poverty, or suggest that poor people accept their poverty. We cannot ignore the need for a fairer distribution of resources within and between the nations of the world. Nevertheless, in our view, virtues are virtues because they further flourishing in the world as it actually exists, warts and all, or because they help us improve it. Individuals may choose to fight injustice, or focus on furthering their own well-being within particular unjust parameters; they may strive to rise out of poverty, or make their peace with it. In all these cases, we suspect the efforts of those who possess the virtues of simplicity, temperance, and frugality will be more effective than the efforts of those who lack them.

Now consider a sixth and very different kind of objection, from someone who says, "Flourishing be damned! I'll be happy if I get to do whatever I want, whenever I want to do it." This might be the two-pack-a-day smoker, or the fat man who insists on his pint of ice cream after dinner. Or consider the strident materialist, who claims, "Simplicity be damned! I'm happiest when I'm buying more stuff, or showing it off to my friends." We *want* to say such objectors aren't acting rationally. How could they risk dying of emphysema, or cutting 25 years off their lives, for such ephemeral or trivial pleasures? How could they value possessions or status more than experiences and achievements, the real stuff of life? But reason's dictates are not completely clear here. Perhaps all we can do, in such cases, is try to get people to distinguish between long-term happiness and short-term pleasure[19]; and between a subjectively happy life and a life of objective accomplishment that they could imagine endorsing on mature reflection (Czikszentmihalyi 2004).

We might also ask these objectors what kinds of character they believe they are cultivating through their consumption decisions. Burroughs (2002) have shown that the self-centered nature of materialism leads individuals to replace intrinsically satisfying personal relations with less fulfilling, extrinsic relations to objects. They have shown that this substitution directly conflicts with both family and community values. What type of people do we want to become? Do we want to be selfish and focused on possessions as central to our satisfaction and well-being? Or, would we rather make personal development, family relationships, community successes, and real hope for a sustainable future our greatest sources of happiness? We are, as Milton Friedman liked to say, "free to choose."

A seventh and final objection might claim that materialism and high levels of consumption are necessary to the proper functioning of society, because they keep the economy growing; hence that simplicity, far from being a virtue, is in fact a vice. While philosophers are not likely to raise this bold objection, it finds voice soon enough on those rare occasions when members of our society debate the idea of limits to growth (Nash 1998). This objection was first memorably raised and subtly defended by Bernard Mandeville, in his "Fable of the Bees." Fully answering it would involve exploring and defending the steady-state economy as advocated by Daly and Cobb (1989) and comparing its long-term prospects to those of our current, endless growth economy. We believe the endless growth economy is an ecological impossibility and a blind alley in the human career, which is destined for higher things. For these reasons, we believe that whatever the obstacles, humanity

[19] Pleasure is ephemeral, experienced most strongly when we satisfy basic needs for food, sex, safety, or comfort. Happiness, on the other hand, refers to the positive feelings people experience when they break through the limits of homeostasis—when they do something that stretches them beyond what they are (Seligman 2000). Psychologists have shown that pleasant experiences are not the sole or even primary determinants of happiness, thus understood. We are all probably aware of some vice we pursue that while pleasurable in the moment, is likely to cause us unhappiness in the long run.

must eventually transition to a steady-state economy. Defending such a large claim is a project for another day. *Creating* such an economy is probably a project for a better generation of Americans. When such an effort is eventually made (probably under duress) the full value of material simplicity may become clear. But in the meantime, we believe simplicity's many benefits to us, right now, amply justify its status as an important human virtue.

Conclusion

We do not mean to dispose of these seven objections too easily. Each of them points to some limits or qualifications to the virtue of simplicity, or suggests complications in fully specifying or living it. But in the end, they do not shake the case for this virtue, because reining in consumption is more necessary than ever to human and nonhuman flourishing. Today one in three children born in the United States is likely to develop diabetes during his or her lifetime, due to poor diet and obesity; the atrophic "dead zone" at the mouth of the Mississippi River is larger than New Jersey, and growing; and global temperatures continue to climb. In a world where all this is true, simplicity is indeed a virtue.

References

Bennett, W. J. (1993). "Quantifying America's Decline". *Wall Street Journal.* March 15

Berry, W. (1990). *What are people for?* New York: North Point Press.

Burroughs, J. E., & Rindfleisch, A. (2002). Materialism and well-being: A conflicting values perspective. *The Journal of Consumer Research, 29*(3), 348–370.

Cafaro, P. J. (1998). Less is more: Economic consumption and the good life. *Philosophy Today, 42*, 26–39.

Cafaro, P. J., Primack, R. B., & Zimdahl, R. L. (2006). The fat of the land: Linking American food overconsumption, obesity, and biodiversity loss. *Journal of Agricultural and Environmental Ethics, 19*, 541–561.

Cafaro, P., & Staples, W. (2009). The environmental argument for reducing immigration into the United States. *Environmental Ethics, 31*, 5–30.

Camarota, S. A., & Kolankiewicz L. (2008). Immigration to the United States and world-wide greenhouse gas emissions. *Center for Immigration Studies*, http://www.cis.org/GreenhouseGasEmissions, Accessed 28 Sept 2008.

Claxton, G. (1994). Involuntary simplicity: Changing dysfunctional habits of consumption. *Environmental Values, 3*, 71–78.

Comte-Sponville, A. (2001). *A small treatise on the great virtues*. New York: Henry Holt.

Crocker, D. A. (1998). Consumption, well-being, and capability. In D. Crocker & T. Linden (Eds.), *Ethics of consumption: The good life, justice, and global stewardship*. Lanham, MD: Rowman & Littlefield.

Cross-National Collaborative Group. (1992). The changing role of depression: Cross-national comparisons. *Journal of the American Medical Association, 268*, 3098–3105.

Czikszentmihalyi, M. (2004). Materialism and the evolution of consciousness. In T. Kasser & A. D. Kanner (Eds.), *Psychology and consumer culture: The struggle for a good life in a materialistic world.* Washington, DC: American Psychological Association.

Daly, H. E., & Cobb, J. B. (1989). *For the common good: Redirecting the economy toward community, the environment, and a sustainable future*. Boston: Beacon Press.

De Graaf, J., Wann, D., & Naylor, T. H. (2002). *Affluenza*. San Francisco: Berrett-Koehler.
Felice, J. (2007). Food for thought. *Acta Neuropsychiatrica, 19*(5), 321–323.
Goleman, D. (1995). *Emotional intelligence: Why it can matter more than I.Q.* New York: Bantam Books.
Heyne, P. (1994). *The economic way of thinking* (7th ed.). New York: MacMillan Press.
House, J. S. (1986). Social support and the quality and quantity of life. In F. M. Andrews (Ed.), *Research on the quality of life*. Ann Arbor, MI: Institute for Social Research.
Hursthouse, R. (1999). *On virtue ethics*. New York: Oxford University Press.
Intergovernmental Panel on Climate Change (IPCC). (2007). *Fourth Assessment Report: Climate Change 2007: Synthesis report: Summary for policymakers*. Access at www.ipcc.org.
Jacka, F. (2007). "Food for thought." *Acta Neuropsychiatrica, 19*(5), 321–323.
Kanner, A. D., & Soule, R. G. (2004). Globalization, corporate culture, and freedom. In T. Kasser & A. D. Kanner (Eds.), *Psychology and consumer culture: The struggle for a good life in a materialistic world*. Washington, DC: American Psychological Association.
Kasser, T. (2002). *The high price of materialism*. Cambridge: MIT Press.
Kasser, T., Ryan, R. M., Couchman, C. E., & Sheldon, K. M. (2004). Materialistic values: Their causes and consequences. In T. Kasser & A. D. Kanner (Eds.), *Psychology and consumer culture: The struggle for a good life in a materialistic world*. Washington, DC: American Psychological Association.
Kottler, J., Montgomery, M., & Shepard, D. (2004). Acquisitive desire: Assessment and treatment. In T. Kasser & A. D. Kanner (Eds.), *Psychology and consumer culture: The struggle for a good life in a materialistic world*. Washington, DC: American Psychological Association.
Lane, R. E. (1998). The road not taken: Friendship, consumerism, and happiness. In D. Crocker & T. Linden (Eds.), *Ethics of consumption: The good life, justice, and global stewardship*. Lanham, MD: Rowman & Littlefield.
Lashof, D., & Hawkins, D. (2006). *An action plan to reduce US global warming pollution*. Washington, DC: Natural Resources Defense Council.
Leopold, A. (1949). *A sand county almanac and sketches here and there*. New York: Oxford University Press.
Lichtenberg, J. (1998). Consuming because others consume. In D. Crocker & T. Linden (Eds.), *Ethics of consumption: The good life, justice, and global stewardship*. Lanham, MD: Rowman & Littlefield.
Maslow, A. H. (1999). *Toward a psychology of being*. New York: John Wiley & Sons.
Millennium Ecosystem Assessment. (2005). *Ecosystems and human well-being: Biodiversity synthesis*. Washington, DC: World Resources Institute.
Mills, S. (2003). *Epicurean simplicity*. Washington, DC: Island Press.
Mokdad, A. H., Ford, E. S., Bowman, B. A., Dietz, W. H., Vinicor, F., Bales, V. S., et al. (2001). Prevalence of obesity, diabetes, and obesity-related health risk factors. *Journal of the American Medical Association, 289*, 70–325.
Nash, J. A. (1998). On the subversive virtue of frugality. In D. Crocker & T. Linden (Eds.), *Ethics of consumption: The good life, justice, and global stewardship*. Lanham, MD: Rowman & Littlefield.
Nussbaum, M. (1993). Non-relative virtue: An Aristotelian approach. In M. Nussbaum & A. Sen (Eds.), *The quality of life*. Oxford: Oxford University Press.
Pacala, S., & Socolow, R. (2004). Stabilization wedges: Solving the climate problem for the next 50 years with current technologies. *Science, 305*, 968–972.
Peterson, C., & Seligman, M. E. P. (2004). *Character strengths and virtues*. New York: Oxford University Press.
Pollan, M. (2007). *The Omnivore's dilemma: A natural history of four meals*. New York: Penguin Books.
Putnam, J., Allshouse, J., & Kantor, L. S. (2002). U.S. per capita food supply trends: More calories, refined carbohydrates, and fats. *FoodReview, 25*(3), 2–15.
Richins, M. L., & Dawson, S. (1992). A consumer values orientation for materialism and its measurement: Scale development and validation. *The Journal of Consumer Research, 19*, 300–320.
Rolston, H., I. I. I. (1988). *Environmental ethics: Duties to and values in the natural world*. Philadelphia: Temple University Press.
Sandler, R. L. (2007). *Character and environment: A virtue-oriented approach to environmental ethics*. New York: Columbia University Press.
Schmidtz, D. (2008). Choosing strategies. In D. Schmidtz (Ed.), *Person, polis, planet: Essays in applied philosophy*. New York: Oxford University Press.
Schor, J. (1992). *The overworked American*. New York: Basic Books.

Schudson, M. (1998). Delectable materialism: Second thoughts on consumer culture. In D. Crocker & T. Linden (Eds.), *Ethics of consumption: The good life, justice, and global stewardship*. Lanham, MD: Rowman & Littlefield.

Schwartz, B. (2004). *The paradox of choice: Why more is less*. San Francisco: HarperCollins.

Segal, J. (1998). Consumer expenditures and the growth of need-required income. In D. Crocker & T. Linden (Eds.), *Ethics of consumption: The good life, justice, and global stewardship*. Lanham, MD: Rowman & Littlefield.

Segal, J. (1999). *Graceful simplicity: Toward a philosophy and politics of simple living*. New York: Henry Holt.

Seligman, M. E. P. (2000). Positive psychology: An introduction. *American Psychologist, 55*(1), 5–14.

Thoreau, H. D. (1971) [1854]. *Walden*. Princeton: Princeton University Press.

Tiberius, V. (2008). *The reflective life: Living wisely with our limits*. Oxford: Oxford University Press.

U.S. Department of Health and Human Services. (2001). *The surgeon general's call to action to prevent and decrease overweight and obesity*. Rockville, MD: Office of the Surgeon General.

Wensveen, L. V. (2000). *Dirty virtues: The emergence of ecological virtue ethics*. Amherst, NY: Prometheus.

Wensveen, L. V. (2001). Attunement: An ecological spin on the virtue of temperance. *Philosophy in the Contemporary World, 8*(2), 67–78.

World Advertising Research Center. (2007). *World advertising trends: Ad spending by country*. Accessed 8 Sept at www.warc.com.

J Agric Environ Ethics (2010) 23:109–128
DOI 10.1007/s10806-009-9183-4

ARTICLES

The Epistemic Demands of Environmental Virtue

Jason Kawall

Accepted: 27 May 2009 / Published online: 14 June 2009

Abstract To lead an environmentally virtuous life requires information—about morality, environmental issues, the impacts of our actions and commitments, our options for alternatives, and so on. On the other hand, we are finite beings with limited time and resources. We cannot feasibly investigate all of our options, and all environmental issues (let alone moral issues, more broadly). In this paper I attempt to provide initial steps towards addressing the epistemic demands of environmental virtue. In the first half of the paper I provide rules of thumb with respect to (1) how to prioritize our investigations into various issues, and (2) what kinds of information we should seek with respect to these issues, and the levels of epistemic justification we ought to attain. In the second half of the paper, I turn to a modified virtue ethics, appealing to the attitudes of virtuous ideal observers to provide characterizations of morally justified and morally non-culpable actions. I then apply these latter concepts in assessing agents, their actions and projects (with respect to environmental virtue), in light of their investigative efforts, and given their particular circumstances.

Keywords Environmental virtue · Epistemology · Ideal observer · Information · Virtue ethics

Introduction

To be a good person, and to behave rightly, requires being informed. Of course merely having a mass of information does not guarantee good character or behavior—one can possess a great deal of information while being largely morally corrupt. But it does seem a necessary condition. For example, to avoid developing

J. Kawall (✉)
Department of Philosophy/Environmental Studies, Colgate University, 13 Oak Dr., Hamilton, NY 13346, USA
e-mail: jkawall@mail.colgate.edu

greed (or acting in a greedy fashion), we need to have some sense of the value of various material goods, and of the costs of obtaining them. What are the environmental impacts of the manufacture of the goods? Have we trapped ourselves in a job and way of life that we regret, simply in order to be able to afford a large house and similar goods? We need an awareness of the costs and benefits of alternative uses of our time and money.

More broadly, to be a good friend will require information about our friends (their needs, their strengths, their foibles, etc.), to be a good environmental citizen will require information concerning the environmental impacts (both positive and negative) of various activities in which we might participate, and so on. In particular, to be a good environmental citizen will likely be especially demanding epistemically, given the wide range and complexity of issues at stake; we will often need to be informed about work from a range of sciences (and other disciplines), even if in popularized, accessible forms. As such, to be good agents we should spend at least some of our time investigating and acquiring morally relevant information. This could include information about major problems facing humanity (including ongoing poverty and hunger, loss of species and habitat, and global climate change), local issues affecting our communities, the labor and environmental practices of companies for whom we work, or whose goods and services we purchase, charitable groups that we can help with time or money, the extent and nature of our own talents, and so on. Sources for such information could vary—from the websites of various environmental groups, to media reports, to one's own fieldwork, to introspection (to know oneself), to books on relevant topics, to lectures or discussions with others, and so on.

On the other hand, to demand a high level of investigation with respect to every decision we make would seem unfeasible. It is not viable for us to read every report released by every environmental organization, even if each could provide us with valuable, relevant information about a host of important issues. Similarly, investigating thoroughly the origins and impacts of every individual food item we purchase, or every alternative action we could perform, while in some ways desirable, is not a plausible option for us. If such standards of information-gathering were treated as necessary for agents to be minimally virtuous, we would likely never act as we continually attempt to become better-informed, thoroughly assessing various goods, options, etc.

In this paper I attempt to take some initial steps towards clarifying the epistemic demands of environmental virtue by considering a set of related questions:

(1) What would constitute an appropriate level of investigation or information-gathering for an agent to pursue in order to act in a minimally virtuous, non-culpable fashion? How should we prioritize issues for investigation?
(2) What kind of information should we pursue, and what degrees of evidence should we seek with respect to this information?
(3) How should we balance these epistemic demands with other our other projects and commitments?

It is implausible to think that there will be a simple rule that can tell us how frequently and in-depth we must investigate our uses of time, effort, and money—

and how to balance this with our other commitments. Much will depend on our individual circumstances, the importance of the decision, the difficulty of investigating, the demandingness of our other projects, and so on. What can plausibly be provided are (1) rough guidelines or rules of thumb, and (2) standards towards which we can strive, even if we lack simple rules to achieve them. In the first half of this paper I present initial rules of thumb with respect to how we should prioritize our investigations, the kinds of information we ought to seek, and the levels of epistemic justification that ought to attach to such information.

In the second half of the paper I propose a set of standards that appeal to the attitudes of fully-informed, virtuous observers towards the character and actions (including information-gathering) of agents. In particular, I will present accounts of morally justified and morally non-culpable actions, and consider how virtuous ideal observers might come to make such assessments, and how these concepts can help us in properly assessing agents with respect to virtuous (or vicious) environmental behaviors. The use of such concepts, as characterized by the proposed modified virtue ethics, allows us to capture our sense that a plausible account of the epistemic demands of environmental virtue must be sensitive to human agents as finite beings, with multiple commitments, limited time and resources, and with each facing distinct circumstances, with varying personal abilities and interests.

Priorities for Information-Gathering

Aristotle draws attention to the possibility of culpable ignorance, though he does not provide extensive guidance as to what qualifies as such:

> They (legislators) also impose corrective treatment on someone who [does a vicious action] in ignorance of some provision of law that he is required to know and that is not hard [to know]. And they impose it in other cases likewise for any other ignorance that seems to be caused by the agent's inattention; they assume it is up to him not to be ignorant, since he controls whether he pays attention.
>
> But presumably his character makes him inattentive. Still, he is himself responsible for having this character, by living carelessly (Aristotle 1985, 1113b35–1114a5)

We might expect matters to be still more complicated now, given our much greater—and easier—access to information about the world, let alone our local *polis*. Beyond this, the issues we face are changing, and often becoming more complex due to our more extensive global interactions, and changing technologies. There is a growing recognition that we must take into account environmental harms of various kinds (whether these be understood in anthropocentric on non-anthropocentric terms); we need to pay attention to a wider range of moral issues. In more contemporary work, Ronald Milo has explored what he refers to as "moral negligence":

> In cases of moral negligence the agent fails to realize, when he should and could, that his act violates his own moral principles. (Milo 1984, 113)

We will here treat "culpable ignorance" and "moral negligence" as capturing the same phenomenon (as does Milo).

In paradigmatic cases of culpable ignorance, we perform morally wrong actions out of ignorance. For example, a person may toss leftover paint into a sewer storm drain out of a lack of awareness of relevant local laws, and of the environmental impacts of such actions. But culpable ignorance can also manifest itself in cases where we perform actions that seem morally acceptable in themselves, but where we neglect what we ought to do. For example, suppose Claire sits drinking her (fairtrade, organic, shade-grown) coffee. In itself this seems morally acceptable. But suppose that she also noticed what looked like thick smoke coming out of her neighbor's home, but fails to investigate, and so fails to become aware of a fire, and also fails to warn and aid her neighbors. She seems culpably ignorant, and fails to do what she should (e.g., investigating and then acting benevolently to aid her neighbors), even though her drinking of coffee does not in itself seem morally wrong.

More broadly, we need to address the question of what standard of investigation and information-gathering must be met in order to be morally justified in our actions, and to avoid culpable forms of ignorance. A plausible starting point here can be drawn from recent, insightful work by Kathie Jenni on vices of inattention. Jenni focuses on instances where we have indications that violations of our moral values are occurring. She suggests (as a "tentative answer") that

> We need to attend (at least) to seeming violations of our moral values in which we are personally implicated, which we have some power to affect, and to which we have been directed by indications that something is amiss. (Jenni 2003, 289)

Still, Jenni worries about situations where an individual can feel overwhelmed by the scope of the suffering and injustice occurring in the world. A too vivid awareness of such facts could be psychologically damaging, or lead an agent to give in to despair. She thus qualifies the kind of awareness of facts that we ought to seek as follows:

> We need to be aware enough of the facts to form personal policies that are true to our values, but not so vividly aware of them that we cannot continue to function. (Jenni 2003, 289)

Jenni holds that we have similar duties to become aware of instances of violations of our moral values even in cases where we are not ourselves implicated, and may have no way of influencing the situation. According to Jenni, while we might be unable to change such situations, there is value in knowing the true moral state of the world. Furthermore, such awareness would reflect a respect for those who suffer, and could well help us to overcome self-absorption.

Still, certain modifications to Jenni's proposal are in order; these are presented in a spirit of continuing to develop her tentative answer.[1] To begin, Jenni only

[1] And I imagine further modifications and additions to what is proposed here will be in order; the issues are difficult, and it might be best to aim at small steps forward.

explicitly addresses cases where we have indications that our moral values are being violated. But this is perhaps too narrow. Suppose we were to hear of a tsunami or earthquake. Likely there would be much suffering, and we would do well to become informed, and to help. Still there does not seem to be a violation of our moral values at stake. To have a *violation* of our moral values would require moral agents who violate these moral values. In the case of earthquakes (and so on) we have instances of morally bad states of affairs, but these do not yet amount to violations of moral values. Thus, as an initial modification, we can consider a broader class of moral issues, where these include any situations where moral improvement (or deterioration) could occur. If we are given an opportunity to help a local environmental group develop an education program for children this seems a morally significant option for us, though presumably no moral violations are occurring.

On the other hand, as it stands, Jenni's proposal may well still be too demanding to be feasible. She suggests that an appropriate level of awareness for those who are mature and emotionally stable (at least for many violations of our values) would involve acquiring visual evidence of potential violations of our moral values—perhaps seeing film of children working in sweatshops, or confronting images of slaughterhouses, etc. But to achieve this level (or an equivalent) of information with respect to all violations of our moral values seems impossible. It seems we would need to find images of every war, every sweatshop, every clear-cut of Amazonian rainforest, and so on. Beyond this, we could well be required to reacquaint ourselves everyday with new images from stripmines, sweatshops, etc., as new violations occur. There would also be thousands of murders, and individual assaults (and abuses of animals) occurring across the world. We cannot investigate all of these violations of our values, and the demands would be still greater if we were to consider all moral issues. Notice that, taken individually, many of these violations would be quite easy to investigate—we could simply spend some time searching the websites of various news groups, non-governmental organizations, and so on. The problem here lies in there being so many violations of our moral values (and other moral issues), and so many potential sources of information; we simply cannot feasibly undertake all of this investigation, even if each particular instance would be relatively easy to pursue.

It might be suggested in response that we only need to investigate those seeming violations of our values where we have indications that something is amiss. And for many of the environmental and other violations of our values, especially those occurring in distant lands, we would have no direct indications, unless we took it upon ourselves to investigate. But this leads us directly back to our original question concerning the nature and extent of the investigations we must undertake to be minimally virtuous. After all, if we embraced such a response, it would be easy for us to avoid most requirements to become informed at all—simply ignore all news sources as far as possible, and keep to the company of similarly ignorant individuals. In this way we could avoid direct indications of violations of our values, or of moral issues, more broadly. Still, this hardly seems plausible. Instead, we would have particularly egregious instances of morally culpable ignorance. We cannot expect individuals to investigate all moral issues in great depth, but nor can they excuse themselves by simply trying to avoid indications that such issues exist.

Jenni herself provides some plausible additional guidance here. She suggests that "as we are more directly implicated in a problematic practice and as we have more power to affect it, demands on our attentiveness increase" (Jenni 2003, 288). With this modification we begin to have some guidance in terms of prioritization for our investigations; we need not simply attempt to be aware equally of all seeming violations of our values (or all moral issues). Still, here we must consider how to understand the extent of our implication in, or ability to affect, a situation. Three possibilities emerge:

(1) Relative to the ability of other people to affect the situation: do we have more or less ability to an influence this situation than others?
(2) Relative to our own ability to affect other similar or more important situations: do we have more or less ability to influence this situation compared to other situations?
(3) Absolutely: do we have a significant ability to influence a situation, regardless of what others can do, and regardless of our ability to influence different situations?

A good case can be made that (when determining how to prioritize issues for investigation) we should hold that the extent of our implication in, or ability to change, a situation should be understood in relation to our implication in or ability to influence other situations, as in (2). With respect to proposal (1), even if others have more ability to influence a situation, if they are not acting on this ability, it seems we should still investigate the issue. On the other hand, even if others have less ability to influence a situation than we do, it could be that they are sufficiently addressing the situation (perhaps by working together), such that our investigations could be better directed elsewhere. What seems more relevant than (2) is the matter of how effectively others are in fact acting to address an issue (or can be expected to act), and we will capture this with an additional rule of thumb to be described below. With respect to (3) it could be that for individuals in dire circumstances, with almost no ability to influence a situation beyond attempting to address their immediate needs, that any *prima facie* duty to investigate issues would be outweighed by other, more pressing concerns. An agent barely subsisting in a refugee camp will presumably need to focus on more pressing issues, rather than investigating issues in which she is barely (if at all) implicated, with almost no ability to influence. That said, there may be issues within the camp (for example) where the individual may be able to influence the situation to an extent that she ought to become informed, even while focusing primarily on other, more pressing matters. On the other hand, a very powerful agent with an ability to effectively contribute to addressing many important issues will still need to decide which issues to prioritize for herself. Intuitively, we might say that when prioritizing issues for investigation, our ability to influence a particular situation (compared to other similarly significant issues) is most important (2), but when considering more broadly whether we ought to be investigating issues or acting on other concerns, our absolute ability or implication is what is most relevant (3).

Still, more is needed. With only this initial modification in hand we would have a greater duty to investigate a trivial violation of our values where we might have a relatively strong ability to influence the situation, rather than a massive violation of our most fundamental values where we are only slightly less able to have an impact. For example, it might suggest we have a stronger obligation to investigate a case where a child over whom we have significant influence may be being somewhat rude to a classmate (in an ongoing fashion), than a case where we have strong indications that a local company is illegally dumping toxic wastes, and we happen to have some influence on an important company figure. Thus, *ceteris paribus*, we must also prioritize cases in accordance with both (1) the apparent severity or significance of the moral issue at stake, and (2) the extent to which attention and action from others [including individuals, governments, NGO's, businesses, etc.] is, or can reasonably expected to be, inadequate relative to a given issue. The second clause captures the intuition that it may be more important to devote our attention to a slightly less severe issue if we can reasonably expect the more severe issue to receive more adequate attention from others to address it, while the less severe issue is receiving inadequate attention. We need to take into account both the severity of issues, and the proportional attention and action that they are receiving.

We should also take into account the epistemic justification that attaches to the indications that a moral issue exists. We cannot expect people to be aware of moral issues if no evidence of the issues is available to them; on the other hand, when there are vast amounts of evidence readily accessible and widely-known in an agent's community, her ignorance becomes much more problematic. With limited time and resources we cannot pursue every lead, every hint of a moral problem or opportunity. We should prioritize those cases where we have stronger evidence that an issue exists—this helps us to avoid wasting our time pursuing mere rumors or hearsay that are ultimately false or misleading. *Ceteris paribus*, then, we ought to focus more attention on cases for which we have stronger evidence an issue exists, relative to the degrees of evidence we could reasonably expect, given the issue at stake. There are situations (especially violations of values) where we can expect cover-ups, and investigators to have difficulty in disseminating information, etc.; as such we ought to consider the degree of evidence that exists, relative to what we might reasonably expect. For example, if we hear rumors of human rights violations occurring in nation with a highly-repressive regime, or hear rumors of abuse within a tightly-knit religious community, this might be all that we can expect initially, and should count as significant under the circumstances. More broadly, once reports of an issue are appearing in reputable news sources (which certainly need not be restricted to the mainstream media), we ought to investigate more than where we simply have rumor or minimal evidential support.

In addition, we also need to take into account the relative difficulty or costs of obtaining information. The easier or less costly it is for us, given our circumstances, to access relevant information with respect to an issue, *ceteris paribus*, the more attention we ought to devote to the issue. Put otherwise, the easier it is for us to gain relevant information about a moral issue, the stronger our obligation to investigate the issue—after all, it is less of a burden upon us, and more reasonably demanded of us. Of course, this is not to excuse us from investigating in cases where severe violations or opportunities to help are potentially at stake and where investigating is difficult. Rather, it is to hold us to a higher standard of information-gathering when it is easy to do so (all else being equal). And of course, different agents, given different circumstances, might experience different costs to obtaining the same information.

Finally, if we are presented with the choice between investigating two (or more) issues that seem broadly equal in the respects discussed so far, we are permitted, *ceteris paribus*, to prioritize for investigation the issue that most interests us, or for which we feel a particular affinity. Thus, if we have two issues of similar importance, receiving similar levels of attention, etc., but where one concerns preserving a local wetlands (which interests us a great deal), while the second is an issue that doesn't particularly interest us in a city hundreds of miles away, we are permitted to give priority to the former issue. But our interests and preferences should be seen as of secondary importance to the other factors mentioned here; while our interests can justify us in giving more attention to a slightly less significant issue than others of which we are aware, we cannot justify ignoring major issues simply because we have more of an interest in some trivial matter.

We can summarize these initial rules of thumb as follows: We ought to prioritize issues for investigation, *ceteris paribus*, in proportion to (1) the extent that we are personally implicated or appear to have an ability to influence a situation, (2) the apparent importance or severity of the issue, and the relative lack of attention and action that it is receiving or can be reasonably expected to receive, (3) the strength of the evidence that there is a moral issue, and (4) the relative ease or low costs of obtaining relevant information for us. With these factors in hand, we can further take into account (5) our own preferences and interests with respect to issues as a further deciding factor, though one that is secondary to the others. Intuitively then, we ought to give top priority to investigating apparently major moral issues that are receiving comparatively little attention, for which we have strong evidence that they exist, where information is easily obtained and in which we are strongly implicated or have a strong ability to influence the situation.

It is worth noting that as our investigations proceed, our evidential status with respect to an issue may change, and we ought to adjust our investigations accordingly. For example, we might find that a significant moral issue that we reasonably expected to be adequately addressed by others is not being so addressed: if so, our obligation to investigate (and ultimately act) becomes stronger. Or we may initially hear rumors that some large electronics producer is engaged in particularly environmentally irresponsible behaviors with respect to toxic wastes. If, as we investigate, we find that it is a company in which we hold stock, this might require us to pursue the issue more than we initially expected (insofar as we are more implicated than we initially supposed); on the other hand, if the story turns out to be rumor, or we find that the company only sells goods in other countries, that others are taking effective action against the company, and the violations are quite minor compared to other situations of which we are aware, etc., we might find that we have less demanding requirements for further investigation. Our time could be better spent pursuing other projects, or perhaps investigating other, more pressing moral issues.

What Information?

Thus far we have focused only on how we should prioritize issues for investigation. As it stands, we have not yet begun to assess what kinds of information we should be seeking, nor what would qualify as enough information (i.e., once we have begun investigating a given issue). In addition, we also need to determine what level of evidence is required with respect to the apparent data we gather. Suppose we were trying to gain information about deforestation rates in Amazonia. Simply accepting whatever claims are made on some randomly-chosen website or taken from pamphlets published by a dubious organization would hardly satisfy the epistemic demands placed upon us as we strive to lead virtuous lives; such an approach would likely lead to false beliefs and error. On the other hand, it would be excessive to require us to seek out all available potential sources of information concerning Amazonian deforestation, in hopes of achieving epistemic certainty with respect to the issues at hand.

With respect to the kinds of information that we ought to seek when investigating a particular issue, it seems that the following, at least, would be relevant:

(i) Information about the extent of the harms and benefits of what is actually occurring.
(ii) Information about the causes of the relevant states of affairs (including the extent and nature of one's personal implication in them).
(iii) Information about potential alternative actions or policies (including one's options for personal action), both with respect to the issue itself, and with respect to alternative issues or projects.

More broadly, and relevant to most all issues, we would need information of the following kind:

(iv) Moral knowledge/information: this would include at least knowledge of moral rules of thumb, of moral exemplars, and of the actions typical of various virtues.[2]
(v) Information concerning logic and deliberative rationality, and an ability to apply this information. In order to effectively draw upon the other forms of information noted here, an agent must be able to engage in rational reflection and assessment.
(vi) Information concerning human and non-human flourishing; what kinds of activities contribute to human well-being? What helps species (and individuals of these species) to flourish? This would include information concerning local ecosystems, as well as more general information about the Earth's systems and species.
(vii) Information concerning our own individual talents, preferences, resources, and so on. Such information would help us to see what projects are best for each of us to pursue, and help us to determine how to distribute our resources (time, money, effort) among our projects.

[2] See, for example, the discussion of '*v*-rules' in Sandler (2007).

(viii) Information about a wide range of (apparent) moral issues and other potential projects. In order to be able effectively assess any given issue, we will typically need to have some sense of other issues; we need to have information concerning whether there are more important issues, or issues where we are more able to act effectively to address the issue, and so on. Such information will help us in determining which personal projects to pursue, and how to balance these with commitments to address various moral issues.

There may well be other kinds of information that would be helpful to us determining how to choose and balance our long-term projects and commitments, how best to act with respect to various moral issues and so on, but certainly the above list captures several relevant fields of information.

Of course, we still require a standard for how much information we ought to seek with respect to an issue, and to what degree of epistemic justification we ought to aspire. I would propose the following as an intuitively plausible standard:

> Ideally, relative to a given issue, we ought to acquire at least enough information that is sufficiently well epistemically justified for us, to make acting on that information (with respect to this issue) rational for us, given our circumstances.

Put otherwise: we should, ideally, seek to be at least sufficiently well-informed concerning a given issue to be able to act rationally with respect to it, given our circumstances.[3] If we were trying to decide how best to reduce our carbon footprint, we would need to investigate various methods of doing so, learning about what factors contribute most to the carbon footprint of someone in our circumstances, etc. We would also need to find reputable sources for our investigations to ensure that our beliefs are sufficiently justified. Of course, much of our knowledge is grounded in the testimony of others, including other individual agents, but also various groups and organizations. Thus we can—and most often do—acquire information concerning various environmental issues, and how best to act with respect to them, from NGO's, various governmental bodies, and so on (where we have good reason to believe that these sources are generally reliable).[4] On the other hand, we would not be required to amass data on every tiny component of our carbon footprint (for example), nor would we be required to go into such great depth in our investigations as to achieve absolute epistemic certainty for our beliefs (which likely would not be possible in cases of this sort). More broadly, acquiring a smaller body of well-justified, highly relevant

[3] Note that acting rationally with respect to an issue could include rationally concluding that other issues are more pressing, or that the costs of trying to help with respect to the given issue are too great. In other words, rationally acting with respect to an issue might involve choosing to devote ourselves to other projects or issues instead. Finally, the current proposal does not hold that being well-informed guarantees rational action; rather, only if we reason properly, while being so informed, will we arrive at rational results.

[4] For more on testimony as a source of knowledge see, for example, Coady (1995) and Lackey (2008).

information will typically be more efficient than hobbling together larger quantities of less relevant, and less well-justified information.

Note that in some cases we might require more, and better justified information in order to make our actions rational (assuming that we also reason well, of course). For example, if a state were using nuclear power and decision-makers were attempting to determine how to store wastes, given the significant risks at stake, a great deal of well-justified, relevant information would be required in order to allow for rational decisions and actions. On the other hand, in some cases, less information (with weaker justification) would be required in order make action upon it rational. For example, a doctor working in a refugee camp may not have access to all of the diagnostic tools, etc., that she would have in a well-funded hospital under normal conditions. In the hospital more information would be required to allow the doctor to act rationally with respect to a patient (given the circumstances) than in a chaotic field situation with limited resources, and thousands of desperate patients.

The case of the doctor raises the issue of whether it can be inappropriate for an agent, given limited time and resources, to seek further information. While a full discussion of this issue is beyond the scope of this paper, we can tentatively propose the following:

> *Ceteris Paribus*, with respect to seeking further information concerning an issue, to the extent that an agent has good reason to believe that the costs associated with seeking this additional information would outweigh the benefits of acquiring such information (in terms of an improved ability to choose and act rationally concerning the issue), given her particular circumstances, the agent ought not to pursue this additional information.

Under such circumstances, other things being equal, the agent should instead devote her efforts towards other commitments or to acting on the issue at stake given the information she currently possesses, depending on her particular circumstances. In the case of a doctor in a triage situation, it will often occur that while she could gain some further diagnostic information that may be of some use in helping her to treat a given patient, the costs in terms of effort and lost time that could be devoted to treating this and other patients are such that she ought not to pursue the additional information.

Virtuous Ideal Observers

While we now have greater guidance with respect to (1) which issues we ought to prioritize, and (2) the extent and nature of the information we *ideally* ought to pursue in any given case, it still seems that the sheer amount of information-gathering required of us would be overwhelming. Ultimately, even if of lower priority, it seems we would need to investigate almost all potential moral issues, even apparently trivial ones. And if so, we would be trapped in place, doing nothing but seeking awareness and information, without ever actually being in a position to act. We need to have a standard for balancing time spent investigating with time spent actually acting on projects. While it is important for us to become informed, it is presumably still more important for us to act, to lead our lives.

Recall the above proposal that, ideally, with respect to a given issue, we ought to seek at least enough information (that is sufficiently well-justified) to allow action upon it, with respect to this issue, and given the agent's circumstances, to be rational. As it stands, we do not yet have an account of what would determine whether a given level of information is sufficient to allow for rational action. Simply put: who or what determines what would count as sufficient information for rational action? I would propose that this can be understood in terms of the attitudes of virtuous ideal observers. Intuitively, we can look to the approvals of (hypothetical) virtuous beings who are fully-informed about the state of the world; would they suitably approve of the amount and kinds of information gathered by an agent, and the degree of epistemic justification her beliefs possess? If so, the agent's information is sufficient to allow for rational action. More precisely, consider the following:

> *IRR* (Information and Relative Rationality): An agent has acquired at least enough information (that is sufficiently epistemically well-justified) to make acting on that information rational, relative to a given issue and given her circumstances, iff an unimpaired, omniscient virtuous observer would deem this to be so (where such deeming is a form of approval).

It is the approvals (or not) of such virtuous ideal observers that determine whether an agent has acquired sufficient information. Why appeal to the judgments of such hypothetical virtuous beings? First, insofar as they are attributed omniscience, they would know the extent of any harms or sufferings that are occurring (thus also knowing which are most pressing or severe), what any given agent's abilities and opportunities for action are, what actions or policies would most effectively ameliorate any given situation, the relative difficulty for any agent of attaining relevant information, and similar such relevant information. The virtuous ideal observers would thus have the informational resources to make what we would intuitively consider to be appropriate judgments. Second, insofar as they are virtuous, they would be honest and consistent in assessing agents and their actions, would appreciate personal relationships and other commitments, would be benevolent (thus not only would they know of the suffering of sentient creatures, they would care about such suffering), and so on. They would thus have ideal character traits to make what we would intuitively consider to be appropriate judgments; they have character traits that we would wish to find in those making moral assessments.

Notice that an agent's circumstances will presumably be relevant to an ideal, virtuous observer's assessments here—we are considering a standard of relative rationality, one relative to an agent's circumstances and the issue being considered. In trying to decide whether to slowly walk away from a bear or to stand in place, attempting to look large, there might not be time for a thorough assessment of large amounts of information, but a quick appraisal of the situation might be sufficient to be deemed to allow for rational action by a virtuous observer. On the other hand, a wealthy person trying to decide how to allocate her charitable donations—and the extent of these donations, with much available time (the issue being considered isn't immediately pressing) and many resources, might be required to obtain much more information, with much greater evidence, in order to arrive at the best use of the funding that she can provide. Similarly greater informational demands would apply in a case where an agent is attempting to determine how best to reduce her own ecological footprint—or what political/social actions might be most effective in encouraging necessary social changes to reduce national or global ecological footprints.

With a standard established for informational requirements (to be sufficient to allow for rational action with respect to an issue for a given agent), we can turn to a standard of morally justified action:

> *MJ*: An action is morally justified for an agent in a given set of circumstances iff an unimpaired, omniscient virtuous observer would deem the action to be morally justified (where such deeming is a form of approval).

Of course, even if an agent has adequate information (given her circumstances) about an issue to be rational in acting, and even if she does in fact act rationally, this does not guarantee that she will act in a moral fashion. She might ignore or downplay information in a biased fashion even while possessing this information, or she might act to achieve immoral goals in a rational fashion (consider an intelligent Nazi). So a distinct assessment needs to be made: is the agent acting in what would be a morally justified fashion, based on appropriate reasoning and information? Note that a morally justified action may not be a right action (and vice versa).[5] Suppose we gather a great deal of evidence from reliable sources that a certain farm engages in humane animal husbandry, but it turns out that we have been deceived, and that the farm carefully hides quite terrible animal welfare and broader environmental violations. In such a case, our purchasing goods from the farm would have been justified—given the strong evidence we had acquired, this would have been an appropriate decision—even if our action would not ultimately be right. We would have made effective, good faith efforts to do the right thing, so our actions would still be justified.

How are informational requirements attached to moral justification? A necessary (but not sufficient) condition for an action to be deemed morally justified by a virtuous ideal observer is that the agent be acting on the basis of information that meets the requirements of IRR above. Without such information, an agent is, to some extent, merely guessing. In some cases, an agent may simply be unable to acquire the requisite information, even though she must act. In such cases, while her actions might be non-culpable (assuming that the relevant information was inaccessible, or too costly to acquire), this is not yet to hold that she would be justified in her action.

[5] We can characterize right action in a similar fashion: An action is right for an agent in a given set of circumstances iff an unimpaired, omniscient virtuous observer would deem the action to be right (where such deeming is a certain form of approval). Presumably people can do the right thing for the wrong reason (perhaps merely to impress others), or by chance (guessing at the best charity to give to). Better, of course, is to perform morally justified right actions. See Kawall (2002) for further defense of this account of right action.

Still, even if an agent fails to act in a morally justified fashion, this does not yet show that the agent would be blameworthy in her actions. Just as we have characterized (1) informational requirements for relative rationality and (2) moral justification in terms of the attitudes of virtuous ideal observers, we can do so with non-culpable actions:

> *NCA*: An agent is non-culpable for performing an action in a given set of circumstances iff an unimpaired, omniscient virtuous observer would deem the agent to be non-culpable (where such deeming is a certain form of approval).

Such actions, while perhaps not right or justified, are not yet culpable. There are situations where, for example, an agent simply cannot devote enough time to adequately think through an issue to perform a justified action. There will also be cases where an agent must act, but crucial information—information required in order to make a properly informed judgment—is physically unavailable, or only available at far too great a cost, given the issue at stake, and the agent's circumstances. In many such cases her actions will be non-culpable. Note further that all justified actions will be non-culpable. After all, in the case of justified action an agent is appropriately investigating, reasoning, and acting, given her circumstances.[6]

We can distinguish between the morally right, the morally justified, and the non-culpable as follows. Morally right actions are those that are in fact permissible, and those that we truly ought to perform (intuitively, they are actions that are *objectively* right). Morally justified actions are those that an agent performs when she has gathered appropriate information (sufficient to act rationally with respect to an issue), and acts from morally good motives. If all goes well, a morally justified action will typically also be a morally right action, though as noted above, the two can come apart (as when an agent's information is misleading). Finally, morally non-culpable actions are those for which an agent lacks appropriate information in order to act rationally with respect to an issue, but where this lack of information would be excused by virtuous ideal observers.

To better see the difference between justified and merely non-culpable actions, compare two cases. In the first, an agent spends a few hours here and there learning about a local environmental issue, and begins to take some steps towards action (perhaps initially signing petitions, etc.); still, he does not immerse himself in the issue (there are several other people who better-informed than he is), and he still devotes himself more to other worthwhile projects than to this one. Such actions (with respect to the local environmental issue) would likely be justified, given the agent's other commitments and projects; he has at least enough information to act rationally with respect to the issue. On the other hand, consider another agent who lives in the same area, but who struggles with keeping two full-time minimum wages jobs simply to maintain a near poverty-line existence, with two young

[6] What would instead lead agents to culpable ignorance (and ensuing actions)? Milo distinguishes four common forms or sources of such moral negligence: (1) impulsiveness, (2) carelessness, (3) recklessness, and (4) self-deception. See Milo (1984, Chap. 4).

children, and an ill parent. We could well imagine that this person has almost no time that is not taken up with important projects, such that she is simply exhausted during any free time she has, and never investigates (or even hears of) the same environmental issue. In such a case, we would presumably see the agent's lack of information and action as excused, and non-culpable. Intuitively, the first agent has taken steps to become informed in a reasonable fashion, given his circumstances, and acts in a morally justified fashion; the second agent is quite uninformed, and takes steps that are inadequate to become informed or to act on the issue (indeed, she may take no steps at all), *but* given her circumstances, her actions would be excused and non-culpable. Of course, she is also not flourishing as an individual, and may well be having detrimental environmental impacts, even if this is not her fault. Here broader social and political forces are at work, which make it extremely difficult for many of us to lead the best kinds of lives.[7]

We should also distinguish among different cases of people working such demanding hours, etc. In particular, we can distinguish between (1) cases where the agent, if she were to have greater available time and resources, would in fact make appropriate efforts to become informed, to act, and so on, and (2) cases where the agent would, with greater available time and resources, still fail to appropriately seek information, act, and so on. The first agent would be a virtuous person, one who is essentially trapped by circumstances such that she cannot act in the morally justified ways that she wishes (but would so act, given different circumstances). The latter agent would be a vicious person, who faces similar difficulties given her circumstances, but whose character is such that she would act in morally culpable ways, even given more favorable circumstances.

More broadly, note that even if an agent performs a wrong action, this does not yet show that the agent is a bad or vicious person. Particularly, if the person's action is morally justified, we can still have a positive appraisal of her and her action; she has done what can reasonably expected of her from a moral point of view, even if in this case she is unlucky and performs a wrong action. In the case of non-culpable actions, even if they are wrong, we cannot yet conclude that the agent is vicious. Rather, with such actions there are excusing conditions for the agent, such that the agent is not blameworthy for performing the wrong action, and we do not yet have grounds for assessing her character as flawed.[8]

[7] Of course this is not to hold that all individuals struggling under difficult economic circumstances will be ill-informed about broader moral issues, etc. Far from it—leading voices in environmental justice movements (and others) have faced such circumstances. Indeed, living in such circumstances might allow one to be especially well-informed about a range of moral/environmental issues that are undetected by others. But it is to recognize that those who are struggling simply to maintain a modest living will often have fewer resources (including time and energy for effort) to devote to seeking information, and for acting on various moral issues.

[8] On the other hand, it is possible for vicious agents to perform right actions on the current account—they can perform actions that virtuous ideal observers would want to be performed in the circumstances. But this does not yet show the agent to be virtuous or her action to be justified (or even non-culpable), nor does it show that she is praiseworthy for such actions. She could simply be performing the right action for poor reasons; her action would be unjustified. She might do the right thing in order to impress others, or because it happens to be the easiest thing to do, and so on.

General Applications

Not all virtuous ideal observers would be alike—there are different ways of being virtuous (Schweitzer is not identical to Thoreau; neither is identical to Leopold, and so on); and each could approve of different ways of acting, and of leading a life more broadly. As long as at least one virtuous ideal observer would suitably approve of our actions, they will qualify as justified, right, or non-culpable (depending on the precise form of approval at stake); what matters is that it could be seen as appropriate from a fully-informed perspective. A plausible pluralism is thus embraced: we cannot expect that there is some single, particular balancing of an agent's commitments and actions that she must achieve. Rather, there is a broad range of potentially appropriate ways of life available to her. On the other hand, this does not amount to a simple, implausible relativism—the actions and projects of an agent would need to be suitably approved of by at least one virtuous ideal observer, and obviously many actions and commitments would be deemed wrong by all such observers (even allowing that there is a plurality of ways of being virtuous).

It might appear, initially, that appeals to ideal observers will be of no use to us in practically guiding our affairs—after all, they are characterized as omniscient and virtuous, while we are rather more ordinary. How could we possibly know what they would approve of? This initial appearance is misleading. We can begin to gain a sense of what sorts of actions and lives virtuous ideal observers would approve of by looking to the lives of actual virtuous persons. In particular, we can gain a sense of how we should approach information-gathering, and how to balance this activity with others.[9] As we do so, we soon realize that those widely recognized as virtuous environmental agents do not spend the majority of their lives simply becoming informed—their virtue manifests itself in action. Aldo Leopold was a well-informed individual, but he did not investigate (let alone act) upon every environmental crisis that arose; this would simply be beyond the capacities of a finite human being. And a fully-informed virtuous ideal observer would understand such human limitations.

More broadly, we should come to realize that with respect to most moral issues, the best that we can seek is to act in morally non-culpable ways. We do not have the abilities and resources to act in morally justified ways with respect to the vast majority of moral issues occurring around the world; there are likely literally millions of individual violations of our moral/environmental values occurring everyday. Our ideal of seeking enough well-justified information to allow for rational action will need to remain unfulfilled for most issues. This is somewhat disturbing, as merely non-culpable actions will often have harmful consequences—we are merely excused in such cases. Still, we now have initial guidance, given in Sects. 2 and 3, as to how to prioritize issues for investigation, and what information we should seek: *ceteris paribus*, we should prioritize issues to the extent that they

[9] There will, of course, be a possible gap between what would be approved of by actual virtuous agents, and by virtuous ideal observers. Actual agents are not always fully-informed, and may sometimes act out of character. Still, they can function as moral exemplars, and provide a basis for predicting the attitudes of idealized virtuous agents. I defend these claims in greater depth in Kawall (2006). More broadly, (1) we can also appeal to general moral rules of thumb, and (2) our ability to act appropriately will improve as we ourselves become better-informed and develop our own virtues. See Kawall (2003, esp. Sect. 4).

seem of great moral significance, we are strongly implicated or have an ability to make a difference, where evidence is easily accessed.

To the extent that an agent is committed to a manageable set of valuable projects, where she acts effectively, and acquires information satisfying IRR, her actions and investigations with respect to other issues will typically be non-culpable. Thus, for example, imagine a person who (1) is active in local environmental groups, is well-informed about the issues these groups are dealing with, and also (2) has a general awareness of many more severe moral/environmental issues, and acts on these issues, and (3) also devotes time to his family, a range of personal projects, and good friends. Such an agent—like any human—will not have great deals of information about many moral violations and issues, and does not devote time to investigating these issues. But he will likely be morally justified in his actions with respect to those projects and issues to which he is committed, and non-culpable with respect to the issues to which he is unable to devote attention. Given his limited time and resources, further efforts at investigating other moral issues, etc., would likely require taking time and resources away from other projects that would be at least as morally important for this agent. As such, we have good reason to believe that virtuous ideal observers would assess the agent and his actions in the way proposed.

Still, qualifications are needed here. The agent must presumably still pay commonsense levels of attention to developing issues, alternatives and so on (for example, at least getting common information from mainstream media sources). The world is not static, and an agent cannot simply choose a set of worthy environmental projects, immerse himself in them for the rest of his life, ignoring all other possible issues, alternatives, and projects. If the agent starts to gain evidence of moral issues arising that are extremely important, he may need to investigate further, and adjust his actions and projects. If earthquakes occur in a developing nation, causing devastation and massive suffering, and the information is widely known and easily accessible, an agent presumably ought to investigate sufficiently to determine how best to help, and perhaps devote less time or money to one or more of her standing projects for a time. An agent is not fully virtuous if she simply devotes herself to a single, relatively minor moral issue and largely ignores all others.

More broadly, an agent must reflect upon her life from time to time, and consider her projects and commitments, and the alternatives available to her. What was once an unfeasible project may now be viable given an agent's changing circumstances. We may discover issues and projects that speak to us, that we may wish to make space for in our lives. We similarly need to honestly assess our own projects and commitments—do they all contribute to a good, flourishing life? Are we perhaps deceiving ourselves with respect to our interest or ability to pursue a project, or its importance?

There are no easy answers to questions of balancing our various commitments and our need to be well-informed, nor are there answers that will apply simply to all individuals. There are many ways of leading environmentally and morally virtuous lives, with different balances (some focused on fewer projects, but in greater depth, with more relevant knowledge; others covering a wider range of projects, though perhaps with less depth of knowledge or action with respect to each particular issue). We have evidence for this as we consider actual environmentally virtuous agents, and the sorts of lives they lead and approve of. There are many such individuals, with different balances among their commitments, different projects, and so on. To the extent that they are well-informed and virtuous, their judgments and actions will tend to mirror those of virtuous ideal observers, and provide us with guidance for leading virtuous lives of our own.

More broadly, given our limited abilities as individuals to investigate and act upon a wide range of issues, we would do well to encourage changes in our social and political contexts that help us to overcome or compensate for these limitations. The situation in which an agent finds herself can strongly influence her ability to act virtuously. In particular, political and social changes could allow us to progress from non-culpable actions (for which we are merely excused) to morally justified actions (which are far more likely to be right actions). Thus,

(i) We should support the work of various non-governmental organizations that can investigate issues for us, and provide guidance as to how best to act. Intuitively, by supporting (through time or money) responsible environmental groups, they can undertake needed research that can in turn be brought back to ordinary individuals; in essence we can delegate some investigation to these groups, helping us to become suitably informed to act in morally justified ways with much less demanded of us. It allows us to act in morally justified ways with respect to a broader range of issues.

(ii) We should encourage greater transparency and provision of information concerning consumer goods. For example, we could encourage retailers or manufacturers to provide more information about how items are produced, so that it is less demanding for agents to make informed decisions. This again would allow us to more easily act in morally justified ways, rather merely non-culpably. If voluntary cooperation is not forthcoming, we could lobby governments to enact legislation to require companies to share such information. NGO's could also play a role here, perhaps producing small, portable guides as to different companies and products, and their environmental record, etc. For example, several groups now produce wallet-sized guides as to which fish species consumers should avoid purchasing. Alternatively, we could look to portable, handheld devices with access to the internet, and having relevant data online. It is especially important to have access to relevant information as decisions are being made.

(iii) We should support greater governmental regulation of damaging industries, etc. Rather than having to carefully investigate the environmental practices of companies with which we wish to interact (which is time-consuming and perhaps beyond the abilities of many of us), governments could impose strict regulations against poor practices—such that we largely no longer need to worry about investigating these companies. If bad practices are not allowed, we will end up acting in a morally right fashion by default—the bad options are no longer present, and no longer need to be investigated (putting aside questions of whether a given item or service is really needed by the individual, etc.). Rather than forcing individuals to try to pick through many options, many of them bad, we can simply regulate against the most dangerous products, etc. Of course some vigilance would still be required—to ensure that regulations are being enforced, for example. But certainly the burden of investigating and acting would be greatly reduced (and we could well imagine NGO's taking care of such investigation).

(iv) We should encourage a shift in media organizations to focus on real stories, not entertainment pablum. The situation is rather grim with respect to the mainstream media at this point.[10] Alternatively, we can at least find and support sources that do provide worthwhile information at little cost to us.

(v) We should encourage employers to provide shorter workweeks, living wages, and opportunities such as job-sharing, etc.; these changes could also be approached through government/public policy. With more time outside of the workplace, individuals would be able to become more involved with their families, communities, and (among other things) to devote more time to investigation and action upon issues.

This list is hardly exhaustive, of course. Still, with such changes we would be able to more easily choose and perform morally justified and right (as opposed to merely non-culpable) actions as the burden of investigation upon us would be significantly reduced.

Finally, it might seem that the proposed epistemic demands are too weak—that we can excuse ourselves from being well-informed about important issues simply by immersing ourselves in others. But this would misconstrue the proposal. That we cannot investigate all issues thoroughly does not yet allow us to justify ignoring important issues, instead exclusively pursuing trivial projects or activities. When we consider the lives led by those who are environmentally virtuous, we should recognize that we ought to investigate, and act upon environmental issues far more than most of us currently do.

Acknowledgments I would like to thank Pete Briggs, Ron Sandler, and three anonymous referees for their very helpful comments on an earlier draft of this paper. Thanks also to Ron Sandler, Phil Cafaro, and Richard Haynes for bringing this special issue to fruition.

References

Aristotle. (1985). *Nichomachean ethics*. (trans: Irwin, T.). Indianapolis: Hackett.

Brainard, C. (2008). CNN Cuts Entire Science, Tech Team. Columbia Journalism Review. http://www.cjr.org/the_observatory/cnn_cuts_entire_science_tech_t.php. Accessed 18 March 2009.

Coady, C. (1995). *Testimony: A philosophical investigation*. New York: Oxford University Press.

Jenni, K. (2003). Vices of inattention. *Journal of Applied Philosophy, 20*(3), 279–295.

Kawall, J. (2002). Virtue theory and ideal observers. *Philosophical Studies, 109*(3), 197–222.

Kawall, J. (2003). Reverence for life as a viable environmental virtue. *Environmental Ethics, 25*(4), 339–358.

Kawall, J. (2006). On the moral epistemology of ideal observer theories. *Ethical Theory and Moral Practice, 9*(3), 359–374.

[10] CNN, a leading American "news" channel/organization recently announced that it was cutting its entire science, technology, and environment staff. See Brainard (2008).

Lackey, J. (2008). *Learning from words: Testimony as a source of knowledge*. New York: Oxford University Press.
Milo, R. (1984). *Immorality*. Princeton: Princeton University Press.
Sandler, R. (2007). *Character and environment: A virtue-based approach to environmental ethics*. New York: Columbia University Press.

J Agric Environ Ethics (2010) 23:129–143
DOI 10.1007/s10806-009-9188-z

ARTICLES

Hume's Knave and Nonanthropocentric Virtues

Paul Haught

Accepted: 27 May 2009 / Published online: 12 June 2009

Abstract This essay offers a critical assessment of environmental virtue ethics (EVE). Finding an environmental ethical analogy with Hume's critique of the sensible knave, I argue that EVE is limited in much the same way as morality is on the Humean view. Advocates of nonanthropocentrism will find it difficult to engage those whose virtues comport them to anthropocentrism. Nonetheless, EVE is able to ground confidence in nonanthropocentric virtues by explicating specific key virtues, thereby holding open the possibility of bridging the motivational gap between anthropocentrism and nonanthropocentrism.

Keywords Hume · Virtue · Nonanthropocentrism · Environmental virtue ethics · Pluralism · Moral disagreement · Moral education · Humility · Wonder · Non-eudaimonistic end

In this essay, I revisit the familiar contrast between anthropocentrism and nonanthropocentrism in the context of environmental virtue ethics (EVE). My interest in EVE concerns its potential to motivate and justify concern for the more-than-human world. Environmental ethicists have long described the transition toward nonanthropocentrism as a dramatic shift in worldview (Taylor 1986; Rolston 1988; Norton 1991). The apparent failure of deontological and consequentialist theories to justify this shift—they arguably achieve much success once the shift has been made—has moved some to look to virtue ethics as offering a possible third way. On this view, the normative force of nonanthropocentric ends is to be explicated in terms of the virtues, and in terms of the various strengths of virtue ethics in general, including, but not limited to its ends of human flourishing and the

P. Haught (✉)
Department of Religion and Philosophy, Christian Brothers University,
650 East Parkway South, Memphis, TN 38104, USA
e-mail: phaught@cbu.edu

strengthening of community bonds (Hursthouse 1999; Swanton 2003). Other features of virtue ethics also explain its attractiveness to environmental ethics. Its adaptability to ethical and scientific naturalism, its resistance to codifiability, and its elevation of the significance of emotional intelligence imply that virtue ethics is impressively adequate to contemporary insights in moral psychology and social behavior.[1] Consequently, virtue ethics seems primed to situate human morality within a much broader (more-than-human) natural context. Yet as much as I find this promise energizing—it meshes with my own metaphysical sensibilities—I hesitate when I think of my students, colleagues, and neighbors who just cannot seem to get the force of nonanthropocentrism, that is, who just cannot embrace the view that the more-than-human-world ought to matter for them morally. When I try to understand why I am bothered by this, turning to philosophy for consolation, I find guidance and even a glimmer of hope, if not comfort, in the moral philosophy of David Hume.

In what follows, I bring several features of Hume's moral thought to bear on the practical (and some theoretical) implications for nonanthropocentrism in EVE. Hume is relevant to EVE for a variety of reasons, not least of which is the significant extent to which his own moral philosophy is arguably virtue ethical.[2] For this essay, where the difference between worldviews is under scrutiny, I locate Hume's critique of the so-called "sensible knave" at the center. With the knave, Hume confronts a legitimate threat to his sympathy-based morality in the form of a prudent, calculating, and deceptively amoral agent who exploits the moral virtues of his sympathetic associates. After examining Hume's response to the knave, I argue that his discussion elicits a useful analogy for diagnosing the practical challenge EVE faces when it includes nonanthropocentric ends in its justifications of the virtues. By identifying the practical limitations in making the case for morality to knaves, Hume reveals that the most convincing case for morality lies in realizing the effects of the full range of virtues. It is these that the knave is unfit to enjoy. Nonanthropocentric accounts of EVE find themselves in a similar situation, one that ultimately directs inquiry to the effects of specific virtues in order to underwrite confidence in the nonanthropocentric worldview.

A Dichotomy of Worldviews

There are many senses of *anthropocentrism*, both normative and descriptive. The sense that I employ in this paper refers to the worldview according to which environmental objectives are approved when they benefit human individuals and

[1] One noteworthy sticking point concerns skepticism with respect to *character* (Doris 2002). Speculatively, virtue ethics can still make significant use of virtue language for describing enduring moral motivations traditionally ascribed to character, if this concept is eliminable. Such a response is arguably consistent with Hume's account of virtue.

[2] There is significant divergence on whether and to what extent Hume is a virtue ethicist, and utilitarian and contractarian readings abound. Among those who have drawn insight from Hume on the virtues, Christine Swanton makes use of Humean criteria for determining virtue (Swanton 2003) and more recently has made a compelling case for how to read Hume directly as a virtue theorist (Swanton 2007).

communities. I take this view to be consistent with the idea that in many cases protections ought to be extended to nonhuman animals, species, and ecosystems to the extent that such protections are beneficial to human consumptive, educational, or recreational interests or are otherwise of negligible cost. Thus, according to this view, the more-than-human world is either of instrumental value to humans or its value is undetermined, but anthropocentrists so-described will likely include ecological principles in their long-range plans for sustaining human communities. In this sense, anthropocentrism is a kind of environmental prudentialism. *Nonanthropocentrism*, by contrast, refers to the worldview that extends moral considerability to features of the more-than-human world. By this view, nonhuman animals, species, or ecosystems may have value for their own sakes, and because of that possibility, the choice between protecting them or satisfying human interests is an intelligible one.

In consequentialist and deontological environmental ethics, these worldviews typically inform the ends of *actions*, with the implication that nonanthropocentrism includes consideration of the non-instrumental value of the more-than-human world as a meaningful end of certain actions; anthropocentrism does not. Most virtue ethical approaches, however, distinguish between justifying virtues in terms of their ends and evaluating actions. Actions reflect an agent's degree of success in possessing a virtue. From the agent's perspective, the choice to perform one action rather than another will reflect her sense of which virtues are relevant to the situation (along with other salient facts about her—e.g., her moral maturity, her emotional profile—and her understanding of the context of action). Her action will be the correct one, then, to the extent that it adequately hits the target of the relevant virtues, and even then, depends on her unique capabilities as an agent.[3] Her action will not be correct because it is anthropocentric or nonanthropocentric, or even because it respects the intrinsic value of things (or not). Rather, her action will be correct because it is compassionate, prudent, sincere, friendly, etc. This analysis suggests that EVE may be able to avoid having to reveal a preference for a particular worldview, since actions (and agents) are to be evaluated relative to the virtues rather than value-theoretical ends. But this suggestion fails to consider the limits of virtues for specifying the scope of their application. For example, helpfulness that responds exclusively to the plights of humans is different from a helpfulness that also recommends responding to the plights of nonhumans, even though in both cases helpfulness is substantively identical (i.e., giving assistance). In practice, of course, helpfulness may be fully discharged if the context only requires consideration of human interests, so my point is not that nonanthropocentrism must be apparent in all specific acts of virtue. Rather, when it should be (from the standpoint of the worldview), a nonanthropocentric end will not be explicable specifically from an account of helpfulness.[4] I take it as relatively uncontroversial that EVE can be adapted to anthropocentrism. The goal of a

[3] Ronald Sandler provides an especially clear discussion of action and virtue in chapter four of *Character and Environment* (Sandler 2007).

[4] What the virtue ethical approach can do is use virtues like *helpfulness* to make the nonanthropocentric worldview more intelligible. For example, many people can understand how biological corridors "help" migratory species. That they care to help such species, however, is to actually inhabit the worldview.

sustainable human future is consistent with virtue theoretical ends of human flourishing and strengthening community bonds. I also find unproblematic the idea of nonanthropocentric EVEs, *in the theoretical sense* that virtues can be defined relative to and justified by nonanthropocentric ends. But if I find myself inclined toward nonanthropocentric applications of virtue and my anthropocentric (but helpful) neighbor does not, EVE appears no better prepared than other theoretical approaches to compel a shift in worldview. And as I discuss later, this issue emerges within pluralistic EVEs, too. Of course, demanding such a result from EVE is perhaps to hold it to a standard that other theories cannot meet; that there are no independent justifications for a worldview is an ineliminable feature of moral life after all. It is in light of this situation that Hume's knave becomes relevant.

Hume's tale of the knave is important for EVE for two primary reasons. First, Hume is able to diagnose a significant source of the difficulty (if not failure) of virtue education in the form of preoccupation with a single virtue. In the case of the knave, the virtue is prudence, in the sense of careful self-regard. On Hume's account prudence capitalizes the knave's reason and precludes his ability to appreciate other sorts of virtuous ends. Second, Hume's confirmation of the benefits of possessing a wider set of virtues than the knave suggests that EVE has much to gain by attending to the typical effects of certain virtues. In other words, if Hume's moral agents gain confidence for their moral outlook (which the knave lacks) by their realization of the benefits of virtue, then perhaps the confidence needed to underwrite nonanthropocentrism in EVE can also be located in the typical effects of certain virtues, and importantly, in virtues that do not already rig the game on the side of the nonanthropocentric worldview.

Hume's Sensible Knave and the Hegemony of Prudence

Hume's account of the knave occurs in his *Enquiry Concerning the Principle of Morals*. For Hume, the knave provokes practical questions concerning the aims and intended audience of moral education. Not everyone, it turns out, will be receptive to the virtues. Knaves in particular are constitutionally unfit for moral education owing to their sympathy deficit. Although this strict naturalistic sense of knavery is important for Hume's sympathy-based morality, it is not applicable to the analogy developed in this paper. Trivially, it is true that some persons are simply psychologically or cognitively precluded from moral agency, but this is not what makes the knave interesting. Rather, it is Hume's depiction of a character who is preoccupied with a narrow but rational self-interest, someone who might even pass for prudent, that makes the knave significant. For the knave so-described *unfits himself for morality* by attending only to what is prudent. He cannot see the sense in what the (other) virtues provide for their possessors even if he understands (and sees opportunity in) the fact that they value them.

Internal to Hume's project, the knave represents an important challenge to his sentiment-based theory of virtue because the knave, in the bulk of his actions, appears to possess a trait others in his moral community find praiseworthy, namely his tendency to act prudently in his rational self-interest. The sad news for the knave's moral associates is he is so thoroughly self-interested and rational that he free-rides on their benevolence, helping himself to all of the benefits of their kindness without ever feeling motivated to reciprocate in kind. The main problem the knave poses for Hume concerns his apparent unresponsiveness to sympathy, which for Hume is a necessary condition for morality. Because the knave lacks sympathy, no degree of argument or other means of persuasion can convince him to be moral. Significantly, the knave exposes Hume's moral theory as a whole to a significant vulnerability: that there could be those among us who lack sympathy, and who, protected by *our* morality, deceive us into promoting their own ends. As Hume explains, the knave "in particular incidents, may think that an act of iniquity or infidelity will make a considerable addition to his fortune, without causing any considerable breach in the social union and confederacy. That *honesty is the best policy*, may be a good general rule, but is liable to many exceptions; and he, it may perhaps be thought, conducts himself with most wisdom, who observes the general rule, and takes advantage of all the exceptions" (Hume 1975).

The knave, being sensible, effectively employs his instrumental reason to optimize his pursuit of his self-interested ends. Consequently, it is no argument against the knave to protest that he is acting irrationally; he already has reason on his side doing his selfish work for him. Hume is thereby resigned to "confess that, if a man think that this reasoning [that the knave cannot be persuaded to morality by reason] much requires an answer, it will be a little difficult to find any that will to him appear satisfactory and convincing. If his heart rebel not against such pernicious maxims, if he feel no reluctance to the thoughts of villainy or baseness, he has indeed lost a considerable motive to virtue" (Hume 1975).

One obvious practical question, then, is what can be done about the knave, and how can the rest of us, that is, those of us who are sympathetically motivated to be moral, protect ourselves against such a person? Hume recommends two responses to sensible knavery. The first is that, in a community dominated by knaves, it is best to act like a knave. Otherwise, one risks exploitation by one's knavish peers; moral integrity is a weakness in a society in which generosity and trustworthiness are consistently used against their possessors. Fortunately, Hume believes this first response is probably unnecessary, and he entertains the empirical hypothesis that most human communities are more characteristically *moral* than they are *knavish*. That is, most human communities reflect the widespread activity of sympathy in their moral norms, and for Hume, these norms allow a much greater degree of evaluative richness than prudent motives alone could support. Hume's second response, therefore, is to encourage non-knaves to retrench their commitment to the full spectrum of human values with its normative richness, which the knave (because of his egoistic constitution) is unfit to enjoy.

On Hume's view, then, the knave is seriously missing out on the benefits of morality, and Hume admonishes the knave for denying himself the superior happiness that stems from self-reflection and morality:

> [the sensible knave], if he has any tincture of philosophy, or even common observation and reflection, will discover that they themselves are, in the end, the greatest dupes, and have sacrificed the invaluable enjoyment of character,

> with themselves at least, for the acquisition of worthless toys and gewgaws (Hume 1975).[5]

To be sure, the knave has much in common with the rest of us, particularly an agreeable penchant for pleasure-seeking. But Hume also asks,

> what comparison between the unbought satisfaction of conversation, society, study, even health, and the common beauties of nature, but above all the peaceful reflection on one's own conduct; what comparison, I say between these and the feverish, empty amusements of luxury and expense? These natural pleasures, indeed, are really without price; both because they are below all price in their attainment, and above it in their enjoyment (Hume 1975).

Of course, the knave lacks the ability to appreciate the goods that Hume describes here. So, other than the unlikely possibility that the knave might possess enough "common observation and reflection" to divert himself from his moral shortsightedness, it is probably best to conclude that Hume's rebuke is not really meant for knaves. Rather it is an injunction to the rest of us to be thoughtful, to reflect on the sources of our moral and other pleasures, and to inquire as to the character one needs to achieve such satisfaction. What such reflection will show is that knavish self-interest—and by affiliation, a disproportionate weighting of prudence—are unlikely to engage us with the wealth of other values that tend to show up in relation to our creaturely constitutions. In short, Hume is appealing to his readers to reflect on their standards for virtue and to take notice of how those standards do not all reduce to the pleasure attained by doing well for oneself.

An additional point needs to be made here. In light of Hume's call for reflection, it is important not to embellish Hume's critique with his otherwise optimistic assessment of human nature as generally benevolent. Hume consistently regards prudence—acting in one's best self-interest—as a virtue.[6] Indeed, prudential considerations are instrumental for Hume in explaining how we are able to act upon the perceived advantages of cooperation and adhering to custom (Hume 1978). Nonetheless, we risk knavery, and its corresponding deprivation of the full range of virtuous pleasures, if we reduce all evaluations to standards of prudence.

Anthropocentrism and Environmental Virtue: On the Limits of EVE

By this point it should be apparent that Hume's discussion of the knave bears resemblance to the relation between anthropocentrism and nonanthropocentrism in EVE. Reiterating my earlier qualification, true sensible knavery (sympathy deficient persons) is not the issue here. Rather, Hume's discussion also draws attention to the possibility of a more attenuated knavery that involves preoccupation with a single virtue, specifically prudence. Whereas the true knave is unfit for morality owing to

[5] Annette Baier thinks Hume overstates his case here. Most things we acquire are not, strictly speaking, worthless. "Wealth is not worthless" (Baier 1992).

[6] Hume includes prudence among the natural virtues, valued for reducing the effects of contingency on one's prosperity. Prudence also encourages one to follow local custom (Hume 1978).

his nature, the prudence-dominated agent unfits herself for appreciating the benefits of the wider range of non-prudential virtues by allowing her attachment to self-interested (instrumental) rationality to occlude her appreciation of social and noneudaimonistic goods made possible through possession of other virtues. This kind of situation, it should be noted, is especially plausible in Humeanism, since practical rationality is constrained by more dominant sentiments. But even if one does not endorse Humeanism, the point still holds that an agent could act in ways that are prudently approvable (for oneself and others) yet fail to appreciate the benefits that Hume's otherwise moral agents are able to enjoy as part of their richer morality. Moreover, such a person will present a particular challenge for moral education because of the totalizing effect of her prudence-informed practical rationality. Such a person may even prove to be hopelessly ineducable relative to a fuller set of virtues and thereby fail to gain full entry into the moral worldview. I take this situation to be quite comparable to the one that arises with anthropocentrism/nonanthropocentrism. If the analogy I am offering is strong, anthropocentrism and its accompanying practical rationality occludes the estimable nonanthropocentric effects of virtue. Those who reason well enough about their ends or are dispositionally in near enough proximity may be gradually educated into the full (nonanthropocentric) implications of many virtues, but the bulk of agents who have unfit themselves to appreciate the more-than-human world will have to be persuaded through other means than by appeal to nonanthropocentric values and virtues.

Hume's diagnosis of what constrains the morally virtuous in pressing their case to knaves and the less virtuous is also useful to consider here. One practical limitation the virtuous may confront occurs if their morality is too closely identified with a worldview that few are fit to share. For Hume, this sort of practical constraint appears most starkly in his skepticism about the reach of religious ethics that are too otherworldly and hence unrealistic or too demanding (Hume 1975). (Considering the great extent to which so many people do find these sorts of ends convincing might imply that Hume is just wrong on this one. However, it would explain why religious justifications for morality are equally unconvincing to the religiously uninitiated or disaffected). A second limitation occurs with respect to the skill of the moral educator in communicating her moral norms. In the *Treatise*, Hume develops his account of the "general point of view," a maneuver in moral evaluation necessary for contravening the potential narrowness of first-person sentimental projections of value by fixing moral evaluation to norms held by a wider moral community (Hume 1978). There he also develops his so-called "narrow circle requirement." This evaluative construct limits the expansiveness of the general point of view by designating effective evaluators as those who familiarize themselves with and apply the norms relative to an agent's particular moral community, where community is defined broadly as those with whom one has regular "commerce" (Hume 1978). In stipulating these standards, Hume is not espousing moral relativism so much as he is articulating methods of achieving consensus across varying scales of moral community. As a result, Hume should be read as seeking to accommodate significant flexibility in how norms are applied. On the one hand, Hume's standards caution against crude judgments of highly localized interpretations of virtue: e.g., politeness requires removing one's shoes in their house, even if we do not expect that in ours. On the other hand, when the effects of conduct typically exceed narrow circles through trade and travel (and it's fair to include the effects of military engagement, international policy, global communication, and the economic and ecological impacts of agriculture and industry), moral discourse becomes more dynamic, bringing in more players and viewpoints, and increasing the demand for skill in moral discernment, and for philosophers perhaps, skill in virtue explication. When taken descriptively, Hume's position on evaluation may give us pause to consider that the moral challenges of our age reflect the collapse of so many narrow circles at so many levels of human and environmental interaction: as an ever-increasing global community we are being challenged to find a center amidst varying scales of "commerce" unprecedented in history. Taken normatively, which I think we can (Haught 2006; Sayre-McCord 1996), Hume is enjoining moral evaluators to avoid the risk of too tightly clinging to the narrowness of their own understanding of virtue. In the present context, these standards expose the practical difficulty that a nonanthropocentrist will have in conveying the normative force of her virtues to her anthropocentrist peers. At the same time, however, they reveal the strongest hand the nonanthropocentrist has to play, which is to call out the anthropocentrist's conduct for its impacts that demonstrably reach beyond any one "narrow circle," and by Hume's account, criticism of that conduct is fair game.

Not everyone agrees. In articulating her own Humean approach to EVE, Jennifer Welchman interprets the reach of Humean ethics along exclusively anthropocentric lines. Advocating a stewardship ethic, Welchman urges that moral agents are called to "develop a greater sensitivity to the noneconomic needs, values, and interests that natural entities and environments *serve* if we are to develop enduring commitments to preserve them" (Welchman 1999, emphasis added). Developing this Humean account of enlightened anthropocentrism, Welchman adds:

> If species or entities about which virtually no one cares, has cared, or are ever likely to care can be saved from extinction only by demanding enormous sacrifices of things for which many people have cared, do care, and will continue to care, then we can and perhaps must permit their destruction. As a rule, uncertainty about our understanding of the natural systems involved favors preservation. However, in those instances where both our certainty and the costs are very high, I cannot see why we must accept the costs (Welchman 1999).

Here, Welchman's precautionary rule tempers the sting that nonanthropocentrists might feel in her defense of anthropocentrism; it appears that only in extreme circumstances will the Humean anthropocentrist allow concern for her own species to override concern for the well-being of others. Yet nonanthropocentrists get stung again later—this time more painfully—at the conclusion of her discussion, when she writes, "To those who believe both that we need to embrace nonanthropocentric principles of value if we are ever to respond adequately to human threats to the environment and that theoretical justification of such principles can move us to act upon them, what I have to offer is certainly not the loaf they desire" (Welchman 1999). In light of Welchman's self-imposed confinement of Humean morality to anthropocentrism, it might be unfair to criticize her argument for failing to accommodate the nonanthropocentric viewpoint. Notably, it is Welchman's skepticism that such a position could be motivating (for moral creatures such as ourselves) that underscores the extent to which she argues that Humean style constraints on virtue ascription ought to be recognized.[7]

Where Welchman and I agree is that anthropocentrism impairs the ability of agents to be motivated by nonanthropocentric norms, and it is not a failure of practical rationality. The force of Hume's critique of the knave is that knaves are difficult to dislodge from their dispositions and appeals to what should matter to them can often fail. As a result, Welchman and I also agree that there are limits on what philosophers can expect of EVE. Advocacy of nonanthropocentric virtues through systematic arguments of their beneficial effects for the more-than-human world is unlikely to reach the hearts of anthropocentrists and the even more environmentally vicious. Where I disagree with Welchman is that Hume's moral philosophy implies that anthropocentrism is the only *rational option* consistent with environmental ethics. Ask any nonanthropocentrist, and they would be puzzled by the suggestion that their virtues or their ends are irrational. In part, this is because the rationality at stake here is not merely a practical one, but a theoretical one concerning the relation between virtues, their typical effects, and their justifications. But there are important limits to EVE on this account as well.

To see what they are, it is useful to consider Ronald Sandler's responses to both the practical limitations of EVE and the responsibility of EVE to provide justification for valuing nature nonanthropocentrically. One of the strengths of Sandler's approach is that it explicitly shows how EVE can be inclusive of nonanthropocentrism while acknowledging the diversity of senses in which agent or community ends are also constitutive of virtues. Sandler's pluralistic teleological account of the virtues carves out a range of ends that both determine and are supported by the virtues. Not surprisingly, among these diverse ends are those consistent with anthropocentrism. For Sandler, these agent and community-directed ends fulfill the eudaimonistic function of virtue ethics. Noneudaimonistic ends also appear in Sandler's argument, and he appropriates and refines Paul Taylor's theory of inherent worth to inform and justify the relation between virtues and environmental entities that possess a good of their own. On Sandler's view, this means that environmental collectives such as species and ecosystems lack intrinsic value, but they are nonetheless frequently the proper targets of environmental virtue because of their instrumental relation to individual organisms (and some collectives, such as ant colonies) that also possibly possess goods of their own.

Sandler understands that the pluralistic teleological theory runs into practical limitations. He mentions these as part of his assessment of three adequacy

[7] Welchman's argument also assumes that a Humean theory of human nature, if correct, would preclude nonanthropocentrism from ever gaining normative force. Part of this stems from Welchman's strong internalist reading of Hume on moral value. It is not clear, however, why internalism should necessarily prevent an agent from responding to intuitions of value, especially those associated with the possession of virtues as in Hume's case. Difficulties with his position notwithstanding, Callicott has at least articulated one way in which subjectivism can accommodate the notion of nonanthropocentric intrinsic value (Callicott 1989). For a discussion of the implications of Humean subjectivism for environmental ethics that develops these within the context of Hume's theory of virtue, see Haught (2006).

conditions for environmental ethics in general. The first two assert that an environmental ethic is adequate to the extent that it can justifiably criticize environmentally destructive actions, and that it can guide action and policy, and on both counts Sandler is comfortable that his theory passes the test. The third adequacy condition, however, produces a potential obstacle for Sandler's theory with its demand for practical efficacy. As he explains, an environmental ethic must provide "arguments, reasons, or justifications that are efficacious in moving people to perform the actions or adopt the policies that are recommended" (Sandler 2007). Sandler proceeds to describe the contemporary scene involving American environmentalist attitudes, noting the many instances of gaps between belief (e.g., "environmental protection is good") and motivation (e.g., "I'm not ready to change my consumptive and recreational practices").[8] Despite these practical impediments, even if many Americans fail to appreciate environmental values or lack the relevant virtues, Sandler argues that advocacy for removing obstacles to environmental virtue (e.g., facilitating opportunities for wilderness experiences or promoting environmental awareness education) would nonetheless be warranted. Some of the warrant is diminished since it is unknown to what extent such actions would be effective. However, the motivation for these actions is virtuous, aiming at increasing the prevalence of environmentally virtuous behavior, even if the outcomes of specific educational practices is uncertain.

Is EVE Necessarily Nonanthropocentric?

In explaining his methodology, Sandler distinguishes his pluralistic teleological approach from preceding efforts in EVE. One of these, which he calls the "extensionist" approach, is typified by explicating virtues for their appropriateness for environmental ethical ends. The extensionist approach is weakened, however, by its tendency to ignore justifications for the ends of virtues: "it is not possible to employ the extensionist strategy without a background account of what makes a character trait a virtue" (Sandler 2007). Sandler's approach attempts to remedy this weakness by including nonanthropocentric justifications of the virtues as part of a wider set of justified ends. Specifically, he calls upon Taylor's theory of inherent worth to underwrite a select set of *noneudaimonistic* virtues, what Sandler refers to as virtues of respect for nature, and what makes a virtue of respect for nature a virtue is the extent to which it fits an agent well for respecting the inherent worth of living organisms and some environmental collectives.

For the sake of space, I will not reiterate Sandler's argument for inherent value. Its present importance stems from its resemblance to Hume's injunction to the knave. Citing Taylor, he writes

[8] In qualifying this observation, he notes, "No theory of environmental ethics is going to have an overwhelming effect on people's environmentally related behavior," referring to data indicating that other factors—e.g., behavioral, resource-contingent, and structural—are often greater impediments to moral development than failure to appreciate environmental values (Sandler 2007).

> By [the attitude of respect for nature, Taylor] means that it represents entry into a particular moral outlook, not a position justified by some already legitimated outlook. It is justified instead by a '*belief-system* that underlies and supports the attitude' and shows 'that it is acceptable to all who are rational, factually informed, and have a developed capacity of reality awareness' (Sandler 2007).

If this sounds familiar, it is because Sandler (via Taylor) is echoing Hume's response to knavery, and it likely to have similar results. For nonanthropocentrists, the argument grounds their confidence that the targets of their virtues are indeed rational and responsive to truly worthwhile things. For anthropocentrists, the effect of the argument is less clear, contingent on the extent to which their rationality is depleted already by their anthropocentric commitments. Nonetheless, the argument is very likely the best one available to nonanthropocentrists in support of their virtues, just as the argument for the richness of virtue was for Hume against the knave. Yet as an important caveat, doing virtue theory in this way does not provide an argument for taking up the nonanthropocentric perspective by appealing to practical or theoretical rationality; rather, it assumes the perspective will be convincing to reason once one dwells within it. To the uninitiated, much of the success of the argument depends on effective proselytism rather than garnering rational assent. This circumstance arises because nonanthropocentric virtues imply a capacity to grasp outcomes of modes of discernment that cannot be demonstrated by rational argument. To put it another way, to the anthropocentrist, the claim that the attitude of respect for nature will be acceptable to all who have a developed capacity of reality awareness can easily be turned around: concern for nonhumans over human interests is just as likely to imply reality blindness. The good news for EVE is that virtue ethics does not have to depend exclusively on rational assent. Learning, practicing, and possessing the virtues can all be instrumental in preparing one to dwell within a particular worldview. Once the worldview is taken up, the rational justifications for the virtues fall into place, and it's here that Sandler's defense of the attitude of respect for nature can be deployed.

Interestingly, Sandler's pluralism may also undercut the force of the nonanthropocentric justification for the noneudaimonistic ends. It leaves open the possibility that an agent can switch-off between eudaimonistic and noneudaimonistic ends, since "Virtues of respect for nature do not have special status within the virtue-oriented approach to environmental ethics" (Sandler 2007). Moreover, Sandler explicitly wants to avoid the implication that the attitude of respect for nature is all-encompassing, that it "is the moral attitude" (Sandler 2007). However, I wonder that he is not overstating his need to resist such implications. First, the inherent worth that justifies the virtues of respect for nature is conjunctively included with other determinations of virtue. As Sandler explains, "the inherent worth of living things provides only one end against which character traits are evaluated, the dispositions constitutive of the virtues of respect for nature will be informed *also* by other ends" (Sandler 2007, emphasis added). Second, Sandler's argument for differential compassion—intended to show how an environmentally ethical compassion avoids the practical pitfalls of biocentrism—implies that whatever switching-off is allowed by virtue, it does not permit a choice of targets of compassion: the well-being of individuals, whether human or nonhuman, is always primary (Sandler 2007).[9] Therefore, the pluralism of Sandler's theory does not imply that an environmentally virtuous person must (sometimes) choose between a belief-system that accommodates the inherent worth of things and one that does not. If this is right, then Sandler's view is both consistent with and arguably a gain for nonanthropocentrism. But it also means that Sandler's EVE elevates the importance of nonanthropocentric discernment by fixing the noneudaimonistic ends of virtues to an arguably nonanthropocentric belief-system. EVE allows that virtuous persons will sometimes act in ways that put the interests of persons and communities ahead of nonanthropocentric ends. However, a virtuous person will not be one who shares the anthropocentrist's perspective that environmental responsiveness is limited to the benefits of that conduct for persons and their communities. Moreover, the reason that the anthropocentrist holds the perspective that he does is that, from the perspective of EVE, he lacks the necessary worldview to make the noneudaimonistic ends of the nonanthropocentrist's virtues intelligible.

Sandler's argument is much more comprehensive than I can do justice to here, but in much of *Character and Environment* he struggles to resist the nonanthropocentric implications of his overall account. For example, in explaining environmental decision making, Sandler offers the following anthropocentric rationale for avoiding environmental degradation: "any environmentally virtuous person will be opposed to unnecessary environmental degradation that compromises some peoples' access to environmental goods" (Sandler 2007). However, it does not appear that Sandler intends to restrict environmental virtue only to ends that accommodate human interests, and that the virtuous person is one who also recognizes that "there is nearly always an alternative to some environmentally insensitive behavior or policy that does not compromise the demands of other virtues" (Sandler 2007). Therefore, with Sandler there is an ambivalence—perhaps a necessary one given the demands of cogency for his pluralistic account—over how limited we should understand the reach of the Sandler/Taylor belief-system across the range of virtues to be. Understandably, there are all sorts of circumstances, where environmental considerations are irrelevant to exercising virtue where one would expect the relevant eudaimonistic ones to function. One does not have to be a nonanthropocentrist to grade a test impartially or be attentive to a child's needs. But when environmental considerations are warranted, I read Sandler as arguing that the virtuous person will be inclined to give the more-than-human world due consideration. The rationale for doing so is nonanthropocentric even if "there is no presumption that what an environmental virtue recommends will be subordinate" (Sandler 2007).

If I am correct to tease out this ambivalence in Sandler's account, then it supports my view that EVE duplicates the situation of morality with respect to the knave when EVE aspires to include responsiveness to nonanthropocentric value. As a

[9] Sandler's differential compassion is indeed distinct from what he calls *extensionist compassion*, but both are arguably nonanthropocentric. The difference, as he describes it, is that extensionist compassion supports decisions to intervene to stop the suffering of nonhumans even when the source of the suffering is not anthropogenic (Sandler 2007).

practical matter, the anthropocentrist is not likely be persuaded by appeals to the rationality of nonanthropocentric ends, but nonanthropocentrists may find such arguments reassuring, underwriting confidence in their virtues as they are able to integrate them with a richer moral horizon than their anthropocentric associates are fit to appreciate. There is something troubling about this image, however, in that it reinforces what is perhaps an artificial sense that there is an option between anthropocentrism and nonanthropocentrism. However, with Hume and the knave, knavery is not an option if one wants to be moral. Comparably, anthropocentrism should not be an option if one wishes to be environmentally virtuous. I think Sandler would agree.

EVE and Discernment

Arguably, one of the strengths of EVE is that there is much to be learned about morality from consideration of the virtues themselves (an area where both Welchman and Sandler excel). As an ethical theory, EVE must meet demands of adequacy, applicability, coherence, and consistency, but EVE invites a separate level of reflection on the profiles of specific virtues across a wide sweep of cases. Sandler consistently reminds his reader that the exercise and even excellence of virtue is contextual and agent-relative. As a result, not all that is ethically important about virtue is exhausted at the level of theory, even as pluralistic a theory as Sandler's. The knave's unfitness for virtue also generates the importance of looking to the specific effects of virtues in Hume's theory. By understanding the profile of each virtue, a sense of what the knave is missing may start to become clear. The knave fails to understand generosity, for instance, because he cannot possess the virtue. Non-knaves may be similarly unclear about the virtue, but for them education is still an option. By acting in ways consistent with generosity, they put themselves in position to enjoy its effects. In EVE, explication of relevant virtues would likewise aim to shed light on how they give rise to the nonanthropocentric worldview. This exercise will be less suspicious to anthropocentrists if it does not already assume the validity of a nonanthropocentric worldview. As a theoretical exercise in virtue explication, it will be important to begin with virtues that do not necessarily imply nonanthropocentrism. Thus, the goal is to identify virtues that typically elicit effects that could engage their possessors to shift perspective to a wider, more inclusive moral outlook.

To sketch how this approach might work, consider *humility* and *wonder*. Thomas Hill notably describes humility as supporting a family of other virtues relevant to environmental attitudes, such as cherishing (that which is important to us) and gratitude (Hill 1983). Adding to Hill's considerations, humility's profile highlights the importance of relation, typically a relation of dependence on the part of the humble person to something else. Humility gives leeway to something other than oneself, yielding to its presence, giving it a voice, acknowledging the significance of its otherness. As a result, humility is consistent with nonanthropocentrism by fitting the agent to discern her particular modes of relating to otherness. To be humble is often to put oneself into a proper relation with that which is other and sometimes greater than oneself. To be virtuous, humility cannot refer to self-ingratiation or obsequiousness, but it clearly points in the direction of the value of something more-than-oneself, bringing it close to nonanthropocentrism.

For its part, *wonder* has similar effects for expanding an agent's horizon of concern. In Book II of the *Treatise*, Hume's comprehension of wonder is closely tied to a critical feature of human psychology, namely the love of truth (Hume 1978). Wonder, as curiosity, helps Hume explain why we take pleasure in mathematics and games, the latter of which, though not intrinsically valuable, are pleasing because generally "life is so tiresome a scene" (Hume 1978). More seriously, Hume notes that lacking wonder might make us indifferent to our own reality. For Hume, this implicates wonder strongly in the level of interest one takes in "morals, politics, natural philosophy, and other studies, where we consider not the abstract relations of ideas, but their real connexions and existence" (Hume 1978).

In environmental ethics, Holmes Rolston considers that we may have a duty to "Keep life wonderful!," to sustain scientific, aesthetic, philosophical, religious, and recreational enjoyment of nature as well as to reveal nature's intrinsic value (Rolston 1988). Conversely, wonder might also be attractive within a narrower anthropocentrism, suggesting that wondrous human civilization could substitute for a wondrous more-than-human world. The suspicion lingers, however, that parents and educators have failed if wonder is to be cultivated within limits that confine it to the purely human. If it is not blameworthy to wonder only at what is near and familiar, it is nonetheless consistent to appreciate it for the apparent boundlessness of its effects. Wonder may be responsible as well both for liberating the objects of our desires from our claims on them as ends and for informing us of potential ends for our desires. In making the case for wonder's importance for love, Luce Irigaray claims that the object of wonder remains "impossible to delimit, im-pose, identify (which is not to say lacking identity or borders): the atmosphere, the sky, the sea, the sun" (Irigaray 1984). The value in inculcating a sense of wonder in children is, for Martha Nussbaum, relevant for informing them of potential objects of love and compassion: "Wonder, as non-eudaimonistic as an emotion can be, helps move distant objects within the circle of a person's scheme of ends" (Nussbaum 2001). For his own part, Rolston has been a champion of the view that scientific and aesthetic engagement with the natural environment is not only a consequence of wonder but an activity that enhances wonder, bringing with it increased possibilities for value. In short, if wonder has these effects, not only is wonder thereby important for liberating us from life's "tiresome" scene, but wonder is a virtue especially critical for fitting its possessor to discern the kinds of value that underwrite the nonanthropocentric stance in EVE.

"What are the benefits of morality if I am perfectly content with my own self-interested outlook?" Were Hume's knave so curious, Hume might have a chance to convince him of those benefits. But only by inviting the knave to learn the virtues, getting him to practice them, and ultimately to put him in position to enjoy their effects is Hume likely to have success. Even then, the knave may not warm up to the moral outlook. Such is the predicament of EVE with respect to the nonanthropocentric worldview. However, like Hume, EVE also possesses a resource in explicating specific virtues to account for the values that agents discern in the world. As such, the knave and Hume's response may provide more than an interesting analogy to EVE and its response to anthropocentrism. It may also reinforce a sense that moral philosophy is itself motivated by virtue, which, if properly discerned, fits one to live a life that is richer in value than the alternative.

References

Baier, A. (1992). Artificial virtues and equally sensible non-knaves: A response to Gauthier. *Hume Studies, 18*(2), 429–439.

Callicott, B. (1989). On the intrinsic value of nonhuman species. In B. Callicott (Ed.), *In defense of the land ethic: Essays in environmental philosophy*. Albany, NY: State University of New York Press.

Doris, J. (2002). *Lack of character: Personality and moral behavior*. Cambridge: Cambridge University Press.

Haught, P. (2006). Hume's projectivist legacy for environmental ethics. *Environmental Ethics, 28*(1), 77–96.

Hill, T. (1983). Ideals of human excellences and preserving natural environments. *Environmental Ethics, 5*(3), 211–224.

Hume, D. (1975). An enquiry concerning the principles of morals. In L. A. Selby-Bigge (Ed.), *Hume's enquiries*, P. H. Nidditch (Rev., 3rd ed.). Oxford: Clarendon Press.

Hume, D. (1978). *A treatise of human nature*. L. A. Selby-Bigge (Ed.), P. H. Nidditch (Rev., 2nd ed.). Oxford: Clarendon Press.

Hursthouse, R. (1999). *On virtue ethics*. Oxford: Oxford University Press.

Irigaray, L. (1984). *An ethics of sexual difference*. (C. Burke & G. Gill, Trans.). Ithaca, NY: Cornell University Press.

Norton, B. (1991). *Toward unity among environmentalists*. Oxford: Oxford University Press.

Nussbaum, M. (2001). *Upheavals of thought: The intelligence of emotions*. Cambridge: Cambridge University Press.

Rolston, H., III. (1988). *Environmental ethics: Duties to and values in the natural world*. Philadelphia: Temple University Press.

Sandler, R. (2007). *Character and environment*. New York: Columbia University Press.

Sayre-McCord, G. (1996). Hume and the Bauhaus theory of ethics. *Midwest Studies in Philosophy, 20*, 280–298.

Swanton, C. (2003). *Virtue ethics: A pluralistic view*. Oxford: Oxford University Press.

Swanton, C. (2007). Can Hume be read as a virtue ethicist? *Hume Studies, 33*(1), 91–113.

Taylor, P. (1986). *Respect for nature: A theory of environmental ethics*. Princeton: Princeton University Press.

Welchman, J. (1999). The virtues of stewardship. *Environmental Ethics, 21*(4), 411–423.

J Agric Environ Ethics (2010) 23:145–166
DOI 10.1007/s10806-009-9186-1

ARTICLES

Heideggerian Environmental Virtue Ethics

Christine Swanton

Accepted: 27 May 2009 / Published online: 17 June 2009

Abstract Environmental ethics is apparently caught in a dilemma. We believe in human species partiality as a way of making sense of many of our practices. However as part of our commitment to impartialism in ethics, we arguably should extend the principle of impartiality to other species, in a version of biocentric egalitarianism of the kind advocated by Paul Taylor. According to this view, not only do all entities that possess a good have inherent worth, but they have equal inherent worth, and in particular no species is superior to any other. In this paper, I elaborate a Heideggerian environmental virtue ethics that slips between the horns of the dilemma. Central to this ethics is the relation of "dwelling" and the many virtues of dwelling, according to which the world is seen as "holy" in a variety of ways. This ethics is importantly local in respect of time and place, but also has universalistic aspects. To understand such an ethics, it is necessary to grasp Heidegger's notion of truth as "aleithia" or opening, which enables us to escape the metaphysical dilemmas besetting ethics in the analytic tradition, including standard virtue ethics. Elaborating this notion occupies a large part of the paper.

Keywords Virtue · Virtue ethics · Heidegger · Dwelling · *Aleithia* · Wonder

A Virtue Ethics of Dwelling

Virtue ethics can be schematically defined as a type of moral theory in which virtue, as opposed to consequences, rights, deontological duties, values, is central. Species of virtue ethics will vary according to views on their structure (is virtue ethics necessarily agent-centered, or can a type of virtue ethics be virtue-centered without

C. Swanton (✉)
University of Auckland, Auckland, New Zealand
e-mail: c.swanton@auckland.ac.nz

being agent-centered, for example), and according to views on the nature and basis of virtue.[1] For Aristotle, at the core of virtue is *phronesis* (practical wisdom), for Hume benevolence, for Nietzsche "life affirmation" understood in a depth psychological way, and in a Heideggerian virtue ethics certain interpretative understandings and emotional orientations or attunements to the world as a whole (*Grundstimmungen*.) The theoretical structures and concepts that enable us to understand what makes traits virtues (or vices) can also vary: for Aristotle it is *eudaimonia*, in neo-Aristotelian virtue ethics agent-flourishing, for Hume what is useful and immediately agreeable understood through the moral sentiments, for Nietzsche will to power (in particular its depth psychological distortions), and for Heidegger, dwelling.

This paper outlines many of the theoretical properties and underpinnings of a Heideggerian environmental virtue ethics. It is primarily concerned with the theory of environmental ethics (as opposed to its application) and for good reason in my view. As recent work in virtue ethical environmental ethics for example testifies, the theoretical base for environmental ethics is in a state of ferment and rich development, and until we have an adequate theoretical base, applications may be flawed or even pernicious. Accordingly, this paper attempts to do justice to the (for analytical-minded philosophers) radical nature of Heidegger's philosophy, while at the same time briefly suggesting how the many layers of his thought may be applied in some broad areas of environmental ethics. Necessarily, space limits critical analysis and detailed application in both depth and scope; work that will have to await another occasion.

Why is environmental ethics in a state of theoretical angst? A fundamental reason is that it is apparently caught in a dilemma. Here is the first horn. We believe in human species partiality as a way of making sense of many of our practices. For example, in the 1992 United Nations Conference on Environment and Development the first principle of the Rio Declaration on Environment and Development stated that "human beings are at the center of concerns for sustainable development. They are entitled to a healthy and productive life in harmony with nature" (cited in Sandler 2007). This anthropocentric picture contrasts with the second horn of the dilemma. As part of our commitment to impartialism in ethics we should extend the principle of impartiality to other species in a version of biocentric egalitarianism of the kind advocated by Taylor (1986). According to this view, not only do all entities that possess a good have inherent worth but they have equal inherent worth, and in particular no species is superior to any other.[2]

The horns constitute a dilemma insofar as the first horn is seen as species chauvinism, whereas the second horn is seen as unable to make sense of legitimate species partialism. However, understood in this way, the dilemma contains a presupposition. Both horns are understood as relying on a metaphysics of value according to which entities have value from a single perspective-less, impartial

[1] I discuss these issues in *Virtue Ethics: A Pluralistic View* (2003), and particularly in a future work on the virtue ethics of Hume and Nietzsche.

[2] See Paul Taylor (1986), p. 155. To forestall objections, Taylor proposes priority principles, but these efforts, and species egalitarianism in general, have received criticism (Sandler 2007, pp. 71ff; Schmidtz 1998).

metaphysical view, which assigns value to entities as if there were ethical brute facts, to which our moral judgments correspond or fail to correspond. Either *our* species is seen as superior on such a view, or alternatively all species are equal. Neither view seems satisfactory. What we need is the recognition that an environmental ethics is appropriate to our life form without that view being seen as inappropriately partialistic and species chauvinistic.

According to Sandler (2007), virtue ethics mitigates the problems associated with species egalitarianism by recognizing that environmental ethics must be appropriate to our *human* "life form" (p. 72). Most forms of virtue ethics are Aristotelian. For Aristotle, the essence of human beings is to be rational, and as is well known, rationality proper to human beings for Aristotle is relative to their *human* nature as a biological creature that develops from childhood to maturity and is affected by the changes wrought by old age. All this suggests a rejection of the species impartialism of Taylor. For how else can our "life forms," premised on practices that favor our species, and *its* "life form," be *justified* from the "impartial perspective" of ethics? However, it appears we are then impaled on the first horn of the dilemma; we seem committed to the idea that partialism and anthropocentrism are licensed by unacceptable beliefs in the objectively based superiority of our species.

Sandler believes that a pluralistic virtue ethics according to which there are many bases and modes of moral responsiveness will avoid this horn of the dilemma. He claims that if "appeals are made to intrinsic value they must be to a conception that is not metaphysical, universal, and foundational…it must be 'contextualized,' emerging through attempts to respond to situational demands" (p. 119). Before we can see how this "contextualized" view of value can resolve the above dilemma, we need to see however just how the ideas of virtue and value mesh in this picture. Although on my view virtue ethics rejects what I have called the thesis of Non-Aretaic Value, arguing that standard "List Theory" values or goods such as pleasure are not good or valuable without qualification, unless suitably infused with virtue (Swanton 2003, p. 34). I have also made it clear that rejection of that thesis does not entail that *all* value is dependent on virtue. In particular the value of such natural items as rock formations, trees, and animals is not dependent on their being virtuous or handled virtuously (p. 40). I take it that this is common ground in a sensible environmental virtue ethics. Rolston (2005) is correct to affirm that the value of natural objects such as trees and so forth is independent of virtue.

Like many others, however, Holmes Rolston III misunderstands virtue ethics when he says that, according to it, "excellence of character is what we are after when we preserve those endangered species" (p. 70), making it appear that for virtue ethics trees and so forth cannot have value independent of excellence of character. This misrepresents virtue ethics, for what virtuous agents are concerned about is action and attitudes in line with the *targets* of virtues, such as environmental virtues whose targets involve for example care for the environment. Though virtue ethics can recognize this, can a virtue ethical theory legitimize behavior that apparently differentiates between the value of say plants and humans, without impaling itself on the first horn of the dilemma—species chauvinism?

A way forward is suggested by Sandler's glossing the "demands of the world" as "being sensitive to our forms of life" (p. 75). In this way, "value" is contextualized, in a way appropriate to virtue. But how is the dilemma resolved? Though Sandler, like myself, believes that virtue is a disposition to respond well to what I have called the "demands of the world" in general (understood in virtue ethics as the *targets* of the various virtues), and not just to one's own interests, I now believe that such an ethics (including my own views, 2003) has not advanced far enough from traditional Aristotelianism. In a nutshell we may put the issue as follows. Aristotelian virtue ethics is naturalistic in that ethics is relative to *what* we are: a *human* being. Hume goes further: for him, virtue is not just relative to what we are but also to where we are. It is also relative to what he calls (in relation to justice) the circumstances of a virtue. Heidegger goes further still in his contextualization of "the demands of the world" in relation to virtue. Ethics is not just a matter of attending properly to what and where we are. Our essence is, as he puts it, one of dwelling. Dwelling is a fundamental *relation* between humans and the world as "being in" and "being with." Objectivist ethics in the analytic tradition, including virtue ethics, has not yet embraced Heidegger's metaphysics/ethics of dwelling, which provides a way of understanding *how* we can live by an ethics appropriate to our life form, without appearing "species chauvinistic."

The gist of the argument can be stated here. In traditional ethics there is a neglect of our nature as beings of limited temporality living in a world with culturally based histories (heritage), sense of place, and in general of our being in that world as one who "dwells." Individual human nature is a nature of relation in, with, and to a world where we dwell for a very limited time. For Heidegger, one's rationality has to be the rationality of a being *that dwells*. Without knowing what it is to dwell, we do not know what form of rationality is proper to human beings, or how to understand the human virtues. According to Heidegger, abstract value assignments such as "we humans as a species are superior" or "we are equal (to rats, say)" make no sense.

Why is this so? Such efforts to calculate value suppose a world of values to which rational agents are responsive—a world of values graspable from a so-called impartial perspective, outside a perspective of dwelling. Of central overall importance in his philosophy of dwelling is the idea that it is constituted by an orientation of seeing as holy, which allows us to overcome our "thinking in terms of values and calculation" (Dreyfus 2006, p. 363). It allows us to reject overly detached and disinterested models of ethics. This is, on my understanding, the main point of Heidegger's essay "…Poetically Man Dwells…" (1971a). Here Heidegger discusses a poem of Holderlin's containing this phrase, but discusses it in relation to the full line: "Full of merit, yet poetically, man dwells on this earth." The point is that if we only measure in terms of "merits" in work, production, talents, degree of "rationality," for example, we do not fully or properly dwell; we do not take the poetic "measure." As Heidegger puts it: "Thus it might be that our unpoetic dwelling, its incapacity to take the [poetic] measure [the poetic *logos*, way of seeing and understanding things], derives from a curious excess of frantic measuring and calculating" (p. 228). Efforts to calculate exact quantities of intrinsic value in types of "natural" entity independently of a dwelling perspective or orientation would also fall within this critique. For Heidegger dwelling is "in essence poetic" (p. 228), even if largely unpoetically man "dwells."

What in more detail is a (proper) dwelling orientation? To understand what it is we need to grasp the nature of:

(a) Heidegger's conception of human beings as *Dasein* (Being-there-in-the-world), and
(b) *Dasein's* orientation to that world as one appropriate to *dwelling*.[3]

Consider first (a), the idea of *Dasein*. To understand what it is to be there in the world as a human being, it is necessary to understand what forms of inquiry and understanding of the world are available to, possible for, that kind of being. For example, Heidegger claims, "[emotional] attunement belongs to the being of man."(1995, 63) Human *Dasein* invests the world with certain kinds of significance, proper to *them* as engaged, embodied, agents having a certain nature, and types of involvement in the world.

Heidegger identifies at what he calls the "primordial" level several nested modes of intelligibility or more generally what he calls forms of "disclosure" of the world for human beings: state of mind as understanding and mood (an emotional type of attunement to the world); and discourse, which enables the communication of, and public intelligibility of, emotion and understanding. As he puts it, "every understanding has its mood." "Every state of mind is one in which one understands…and articulates itself with relation to its intelligibility in discourse" (Heidegger 1967, sect. 31, p. 182; sect. 68, p. 385). All these nested modes of intelligibility: understanding, emotional attunement, and public discourse, are essential for conceiving the world as a dwelling place.

Second, consider (b): what is it for Dasein to *dwell*? Fundamentally for Heidegger to dwell is to be at home. Consider the words of Bob Dylan:

How does it feel
How does it feel
To be on your own
With no direction home
Like a complete unknown
Like a rolling stone.

In this chorus of his song "Like a Rolling Stone" Dylan succinctly describes four aspects of not dwelling:

To be without relationships,
To have no sense of home,
To be unknown, forgotten, invisible,
To be a perpetual wanderer, with no roots.

Consider now the *orientation* to the world that is appropriate to *Dasein* as a being that *dwells*. Recall that an orientation consists of nested modes of "primordial" disclosure: states of mind—notably understanding and emotional attunement—that themselves are made possible through discourse, our communicative relation to our fellow human beings. Such orientation "opens up" the world in ways that we

[3] The ethics of dwelling is largely contained in Heidegger's late work, *Poetry, Language, Thought.*

explicate in the next two sections. What then is the fundamental orientation of *dwelling*? This orientation "opens up" the world as "holy." The notion of "holy" is elaborated in "Application of *Aleithia* to an Environmental Virtue Ethics of Dwelling" and "Heidegger's Notion of Wonder," but the basic idea can be presented here. To see the world as holy is to have a fundamental understanding of, and emotional orientation (*Grundstimmung*)[4] to, the world. This orientation is complex and consists of four unified features.

(1) We are open to the *mystery* of the world as a whole (Being). That is we must recognize that being open to the world in a certain way is always at the same time being closed to it—the world "withdraws" in that respect:

> That which shows itself and at the same time withdraws (i.e., our understanding of being) is the essential trait of what we call the mystery. I call the comportment which enables us to keep open to the meaning hidden in technology, *openness to the mystery*. (Heidegger 1975, p. 55)

What Heidegger is claiming here is that without modes of making the world intelligible there is no intentional relation to aspects of the world. We are not open to it. We cannot discourse about, for example, moral properties or even scientific ones: the world as such is merely Being; a Mystery. However, modes of making intelligible are also modes of leaving aspects closed; a mystery. It is not as if there is some "absolute conception of reality" (as Bernard Williams put it) that reveals Being absolutely and entirely in all its facets.

(2) We are open to the mystery of the world as a *dwelling* place, and as a dwelling place the world consists of the "four-fold"; "earth," "sky," "mortals," and "gods." The "gods" are particularly important for instilling a sense of holiness for they are fundamentally icons that are revered, are importantly symbolic in some way, and that help impart a sense of identity and belonging to our dwelling place. This last feature allows us to focus on and reflect on those matters, so significant for dwelling. The "gods" need not be supernatural; they may be culturally significant objects, special items of nature, even sports teams or everyday items.[5] (For an example of a sports team having such status see Young (1999). I shall also use his example.) In dwelling, objects become what Heidegger calls "things" in a rich sense: a thing is not just "an unknown X to which perceptible properties are attached." (Heidegger 1971b, p. 153) Rather it "gathers to itself in its own way earth and sky, divinities and mortals" (ibid.). It becomes rich with cultural and personal meanings. In this sense a thing is a "gathering or assembly" (ibid.).

[4] I discuss and apply Heidegger's notion of fundamental emotional attunement in my "A Challenge to Intellectual Virtue from Moral Virtue: the Case of Universal Love," forthcoming in *Metaphilosophy*.

[5] Indeed some of my examples below will be precisely of this nature to illustrate the point that dwelling involves *relation* between culture, iconic items, and the "purely natural" such as trees: indeed on a dwelling orientation the efforts to draw a sharp and weighty distinction between the natural and the non-natural are misguided. Hence Heideggerian environmental ethics is not restricted to traditional "natural" items, for it is a more holistic approach.

(3) We are open to the world of the four-fold as something to *care* for and as something in which we have confidence that we will be cared for.

Young (2006, p. 378) describes "caring for" (Schonen) as "double-aspected":

> On the one hand, the dweller is the object of caring for, is "preserved from harm and danger...safeguarded ...at peace" (PLT149)...On the other hand the dweller is the subject of caring for: she is the one who "spares and preserves" (*ibid.*) the things of the dwelling place, becomes, as Heidegger puts it, its "shepherd" or "guardian" (P 252, PLT 184).

Only if we see objects such as bridges as "things" in the rich sense explicated above,[6] does our place of dwelling become "near" to us, where it is not the case that "everything gets lumped together into uniform distancelessness" (Heidegger 1971c, p. 166), and only then can we care and feel cared for. Only then are we "at home," and we can "remain, stay in a place"; "be at peace...be brought to peace."(Heidegger 1971b, p. 149)

(4) We are open to the world as a caring and to be cared for four-fold as a *local* dwelling place, local with respect to time (as a world with a heritage), and local with respect to place. Our dwelling place is *our* place, and is thus special to us. This feature is essential to a sense of rootedness, so important for dwelling.

Dwelling orientation then allows for both a universal and a local aspect of dwelling, as is made clear in aspects (1) and (4). It is not just universal in the sense that *everyone* must care for their *own* local four-fold; rather the attitude to the world *as a whole* must be one where it as a whole is seen as a mystery, a source of wonder, and must thus be respected, even if the exigencies of caring are by and large, more local. Hence we as individuals must respect the dwelling places of others and the planet (and indeed the wider universe) as a whole.

In this paper I cannot hope to cover all four aspects of seeing our dwelling-world as holy. I shall concentrate on the idea of truth as *aleithia* that helps us make sense of (1); and on (4), the idea of the local aspect of dwelling. Further complexity is added by a consideration of the emotions constituting the *Grundstimmungen* (fundamental emotional attunements) of the orientation of seeing the world as holy, in these four aspects. These include, centrally, wonder; feeling secure, safe, and "at home"; caring; letting be,[7] passionate devotion to local icons. This latter is an aspect of a nameless virtue described by MacIntyre (1981) thus: "the virtue of having an adequate sense of the traditions to which one belongs or which confront one" a virtue that is not to be identified with "conservative antiquarianism" (p. 207).

---

[6] For a fictional account of a bridge in fourteenth Century England, described in this rich sense, see Follett, *World without End* (2007).

[7] "To let beings be...does not refer to neglect and indifference but rather the opposite. To let be is to engage oneself with beings." Heidegger, "On the Essence of Truth," p. 144, cited in Malpas (2006), p. 270.

To understand more fully a Heideggerian environmental virtue ethics of dwelling in its various aspects, and the nature of the virtues in such an ethics, it is necessary to grasp at a fundamental level Heidegger's view of our dwelling-relationship to the world. To do that in turn it is necessary to understand his notion of truth as *aleithia*. The next section explicates that notion, while the last two apply it to an environmental virtue ethics of dwelling.

The Four-Fold of Truth as *Aleithia*

Truth for Heidegger in its core sense is a relation between meaningfulness, significance, interpretation, or understanding, on the one hand, and the world on the other. Such a relation, for Heidegger, is realist: it reveals the world as it is. Nonetheless, to get "behind" truth as a relation between significance and the world in order to penetrate to the "core of being" (the "*factum brutum*") by means of what Heidegger calls a "pure beholding," makes no sense.

In the analytic tradition, the relation of truth has traditionally been thought of as a relation of correspondence between propositions or sentences (carrying meaning) that represent the world. Heidegger fought against this representational metaphysics dominated by a perceptual model of cognition, arguing in *Being and Time* that our fundamental orientation towards the world is practical. It is one of engagement. One form of engagement is scientific observation of things and their properties, but it should not be thought that this form of engagement is ontologically basic. Taylor (2006) captures Heidegger's basic position by describing the rejected view and Heidegger's alternative:

> The result [of the rejected view] was a picture of the human thinking agent as disengaged, as occupying a sort of protovariant of the "view from nowhere," to use Nagel's suggestive phrase. Heidegger had to struggle against this picture to recover an understanding of the agent as engaged, as embedded in a culture, a form of life, a "world" of involvements... (p. 203)

For Heidegger, the relation of representation presupposes a more "primordial' notion of truth, one in which various forms of significance relations "open up" or "disclose" the world as it really is. These significance relations are not exhausted by propositional knowledge. Comportment in a world of "equipmentality" as Heidegger puts it, fundamental emotional orientations, art, also disclose the world.

What then is disclosure? As David Weberman points out (1996, p. 387): "'disclosure' is introduced as a technical term in *Being and Time*, Section 75 thus: "Disclose" and 'disclosedness' will be used as technical terms in the passages that follow, and shall mean 'to lay open' (*aufschliessen*) and 'the character of having been laid open' (*Aufgeschlossenheit*)." Given that truth is a relation between significance and the world, forms of disclosure are fundamental forms of intelligibility, what Heidegger sometimes calls "horizons" of disclosure. That is how the world is "laid open."

If truth is a relation between significance and the world, truth presupposes comprehending beings giving things significance. Such beings are not to be thought of as subjects who merely represent objects from the perspective of a spectator, but as beings *dwelling amidst* beings, engaged in and with the world. Hence for Heidegger disclosure as "laying open"; "uncovering," or "unconcealment," is an active process of investing the world with significance. It is the terminus of a process in which truth is "wrested from entities"; "entities get snatched out of their hiddenness." (1967, sect. 44, p. 265)

I have outlined basic features of truth as the disclosure of the world (*aleithia*). It turns out, however, that what Heidegger calls the "wresting" of truth from entities by "disclosure" is a remarkably complex process (which he calls a four-fold), befitting a conception of truth as an active process of opening up the world. Understanding these four features makes for a more detailed and rich application of Heidegger's notion of dwelling orientation, some aspects of which are outlined below.

In work subsequent to *Being and Time*, Heidegger summarizes the aspects of this complexity by understanding the disclosive concept of truth as a multiple unity having four aspects, each of which must be understood for a proper application of an ethics of dwelling to environmental ethics. These are called by Heidegger in *Basic Questions of Philosophy: Selected "Problems" of "Logic"* a "four-fold," whose aspects are summarized in section 8, "The space of the four-fold unitary openness" (1994, pp. 18–19). I shall illustrate the four-fold with a basic conception of the disclosure of ethics, which can be understood as presupposed in, or an ingredient of, the dwelling orientation. We might say that for Heidegger caring is at the heart of ethics, and caring requires a fundamentally benevolent attitude to the world, as Hume saw; but as we shall see in the next section, "dwelling thinking" requires more.[8]

Here is the four-fold.

(1) *The "thing" must "be out in the open."*

As we have seen, for Heidegger, something such as a moral property must be "brought out of oblivion," or opened up (disclosed) before any representation can be made of it. In ethics in the analytic tradition this aspect of openness is particularly problematic; moral properties as objective features of the world outside the beliefs or emotions of the agent are seen as queer and inaccessible. Indeed they have been rendered invisible or transformed in subjectivist, nihilist, fictionalist, and non-descriptivist but assertoric views such as that of Timmons (1999). Timmons for example rejects a correspondence theory of truth for ethics in favor of a view according to which the evaluative content of moral judgments "is not a species of representational or descriptive content" (p. 142) but is the revelation of one's stance on something by means of assertion. They have arguably also been rendered invisible in any moral realism that reduces evaluative properties such as virtue and vice to biologically based properties wholly comprehensible within a scientific outlook of evolutionary theory.

[8] "Environmental benevolence" is also seen as a central virtue in environmental ethics (see Geoffrey Frasz (2005), "Benevolence as an Environmental Virtue"). I agree but argue that it should be shaped by a dwelling orientation, an orientation that also shows that many other virtues are central.

Of crucial importance is that for Heidegger there is a plurality of horizons of disclosure disclosing the world. The idea of things being shown up as they *are* does not presuppose that there is only one way of things being disclosed to us. Our understanding of things is not identical to "a cognitive grasp of an essential definition" (Wrathall 2006, p. 255) such as one provided by science, but presupposes attunements or receptivity to the world, which may be of different types. Consider Wrathall's example of the following definitions of gold, purporting to give accounts of its nature:

> Gold is the noblest of the metals.
> Gold is an element with the atomic number 79. (p. 253)

These definitions should not be seen as providing incompatible essential definitions. Rather each presupposes different forms of truth as unconcealment, allowing for things to be shown up in different ways. Thus the idea of things being shown up as they *are* does not presuppose that there is only one way of things being presented to us. All "essential" definitions, encapsulating what is fundamental in our various ways of engaging with the world, are relative to the different backgrounds or "horizons of disclosure."

How can the first aspect of openness, the "*thing*" being "out in the open," be understood relative to the horizon of disclosure of ethics? I suggest that just as Hume believed that moral properties, notably virtue and vice, are "*powers*" to produce the basic moral sentiments or emotions, in beings emotionally constituted as we are, so for Heidegger, basic emotional attunements allow (real) moral properties to be available to us. That is, our emotional capacities secure the intentionality of such properties; moral language can thereby be understood as being about such properties. This is not to say that our grasp of such properties made possible through these attunements has representational accuracy: for that we need other aspects of the fourfold of *aleithia*.

(2) *The thing must be open in a "region between thing and man."*

The *region* of disclosure is the idea that

(a) What is disclosed is an aspect or "region" of the world, e.g., the world of ethics, and
(b) Such a region is but one aspect of the world.

"Being" is a "plenitude" of "regions": other regions or aspects of the world may be temporarily concealed from us, or totally unintelligible to beings such as ourselves.

We discuss each of these features of the region of disclosure in turn. A "region" is disclosed by a *horizon* of disclosure. We may provisionally summarize the notion of horizons of disclosure with the help of a metaphor. Consider the world as like a diamond with facets. For truth as correspondence to take place the facets have to be "lit up" through backgrounds of understanding. The various facets may well require quite different backgrounds. Some of these will light up the relevant facet only dimly; others more richly; indeed "radiantly." The metaphor is somewhat misleading in two respects. First, it suggests that the facets are just regions in the sense of *parts* of the world, so that in the region of ethics, for example, some areas such as trivial actions or self regarding actions are outside the world of ethics altogether. Opposed to this picture is the idea of ethics as "pervasive": no part of the world is intrinsically outside the scope of ethical assessment. Since this is the picture that I and others have defended,[9] I need to show how it is compatible with the second aspect of the fourfold: the idea of ethics as a "region." The idea is this. Though ethics is pervasive it nonetheless opens up the world as a whole in its ethical *aspect*. No part of the world is immune from ethical assessment but ethical assessment is only one aspect of assessment. Similarly, one might claim, poetry is pervasive in the sense that all the world (and not just the romantic, say), is open to poetical treatment. But that treatment cannot be reduced to say, scientific disclosure. Poetic disclosure has its own horizon of disclosure; its own background of understanding. Science too, is inherently able to apply to the world as a whole; it is not that there are some parts of the world where it is out of bounds, such as the religious. Rather the claim is that no *single* mode of disclosure is adequate to the task of disclosing all aspects of the world or to the task of disclosing it sufficiently richly. Each horizon of disclosure is in this sense totalizing but not, as we might say, absolutizing.

The second way in which the metaphor is misleading is that it suggests that the "facets" are disjoint: sharply separated from each other. On the contrary, as McDowell (1995) points out, the "logos" of the practical and the ethical must be integrated with that of science if we are to have a good grasp of reality. The ethics of dwelling for Heidegger requires a "logos" of what he calls holiness, a certain orientation to the world as a dwelling place, but insofar as Hume is correct in claiming that the world of *ethics* is opened up by an attitude of basic benevolence, dwelling thinking requires a sense of holiness infused by benevolence, love, or caring.

A more accurate metaphor then to describe the relation between the various aspects of the world is this: it consists of regions enclosed within a series of concentric circles, at the center of which is science; moving outwards is ethics (more narrowly conceived) and then dwelling. The dwelling orientation is for Heidegger the most comprehensive. Each "region" is pervasive (no part of the world is in principle beyond the reach of, e.g., science or ethics) and each has its own horizon of disclosure. Hence outer regions are not reducible to inner ones. For the horizons of disclosure of the various regions are not reducible to each other, but as McDowell recognized, they must be integrated with each other.

We consider now the claim that ethics (for example) is in a region between thing and *man*. If one thinks of ethics as a horizon of disclosure, the question arises: what is the "region" of ethics? Consider a putative counterexample to the thesis that

[9] See Swanton (2003), chapter 3, section iii; Scheffler (1992). This view is of course compatible with the idea that not all ethical prescriptions and evaluations are important; for example, the need for a considerate caring parent to choose say the broccoli in the garden rather than cabbage for dinner (because her small child had an unfortunate encounter with a green caterpillar in the cabbage the night before) is not one of the weightier ethical decisions.

emotional disclosure is required for the first aspect of openness of the world of ethics.[10] In this alleged counterexample, the "region" of ethics is the solution of coordination problems and consists of a world of beings who solve such problems, but who have no emotions and no experience of pleasure and pain. Such a world would have no room for benevolence as we normally understand it. There are two responses to this view of the world of ethics. First can we intelligibly describe such a world as a world of ethics? On my view, the practices of such a world, if they are intelligible at all would be too far removed from a *human* world of ethics. As we have seen, for Heidegger, horizons of disclosure are relative to modes of intelligibility proper to *human* Dasein and *its* way of being in the world, which is further elaborated in the fourth aspect of the fourfold (below). In short, for Heidegger, ethics is in a region between thing and *man*. If one is not persuaded by the claim that the world described in the alleged counterexample is unintelligible, it would still be true that on Heidegger's view, to think of the world of ethics merely as the solution of coordination problems is to see that world in a very attenuated and dim way.

(3) *Humans must be open to the thing*

As far as ethics is concerned, the first aspect of openness will be portrayed as requiring the basic moral sentiments, but this is insufficient for openness in this much more demanding sense. In ethics this aspect of openness requires a moral sensitivity, such as Iris Murdoch's attention, Aristotle's practical wisdom, moral imagination, and most importantly for Heidegger, a continuing attitude of questioning and wonder. Individuals who have these features can be seen as authoritative inasmuch as they are particularly "open to the thing."

The distinction between aspect (1) and aspect (3) of openness resolves what I consider to be an ambiguity in Heidegger's idea of disclosures revealing the world "as it really is." The first sense is the idea of "opening up" from "concealment." Without our structuring the world through *logos*, or engaging in it through "mood," there would be no world for us. Being would remain concealed. However not all disclosures in the above sense "open up" equally successfully or accurately. For this to occur, individuals need properties such as expertise, competence, insight, sensitivity, creativity, love, practical wisdom, and so on.

In the second sense of revealing the world as it "really is," disclosures, which reveal the world in the first sense, are relatively undistorting. They do not for example, "veil" to use Heidegger's term. For example, art, or at least good art, discloses the world "as it really is" but more or less radiantly, hammering nails discloses a real world of equipmentality in which hammering with nails is the way to build, but skilful hammering with the right nails discloses that real world successfully or competently. Benevolence, caring, love, disclose the real world of ethics, but true propositions about virtue and vice provide accurate representations of moral properties whereas inexpert but reasonably benevolent agents may provide less accurate ones. Science discloses the world as it really is too, but if it is what

[10] I thank John Doris for this "counterexample" and discussion of it.

Heidegger calls "enframing" it closes off other forms of disclosure, for example the ethical, and may be, as Heidegger puts it, "dry" and "monotonous" (1997, p. 17)

Thus truth as *aleithia* admits of degree, and has significant normative dimensions. For example, a mode of disclosure can disclose the world relatively richly or radiantly, or relatively dimly. This feature, as we shall see when we discuss Heidegger's notion of wonder, is of particular importance for the ethics of dwelling. A horizon of disclosure can also disclose more or less successfully or competently, more or less authentically, and can close off or not close off other modes of disclosure.

Note that on Heidegger's four-fold "space" of openness authoritative individuals who are particularly "open to the thing" are not authoritative in isolation from social contexts of dialogue and negotiation within institutional practices. There is no "virtuous agent as oracle" model here.[11] Note also that moral truth is not necessarily fixed by authoritative individuals, no matter how virtuous and wise, and no matter how well they are socially situated. For Heidegger, such an assumption is to violate in a particularly significant way, this third aspect of openness. For in order to remain "open to the thing," agents no matter how expert or virtuous, must maintain a continuing attitude of questioning and wonder, which of course presupposes a sense of their fallibility.

(4) *Humans must be "open to their fellows."*

As we have seen, for Heidegger, representation can occur only if the world is open to us (truth as *aleithia*) and the world is open to us only if we are *engaged* with it. For us to be engaged with the world humans need to be "open to their *fellows*." We need interpretative understanding. This aspect of openness as interpretative understanding is what has been described as the hermeneutical turn. If engagement is the "primordial" mode of orientation to the world, being in the world (*Dasein*) as Heidegger puts it, then understanding is at the foundation of philosophy. As Hoy (2006, p. 177) claims, "Heidegger makes interpretive understanding the central mode of human existence."

Interpretative understanding requires what Heidegger calls the everyday averageness of *Das Man*, including our language, conventions, and institutional rules and roles. For the "thing" to be accessible, or "out in the open" there must be a background of human practices within which the thing is *related* to other things. Assume that the disclosure of the world of ethics in its first aspect (being out in the open), requires what we shall call "totalizing" moral sentiments (such as benevolence) that enable things such as animals or children to be "out in the open" *ethically as individuals*. These sorts of things, unlike perhaps grains of sand or individual dead leaves, can be seen as having what Hume calls "powers" to produce these sentiments. This, however, is not enough. For the world of ethics to be disclosed, the objects understood as having ethical significance must be *related* in ethically significant ways, comprising the multifarious "involvements" of ethics; educational, political, institutional, family, friendship, and so on. The thing or state

[11] For criticism of this model of virtue ethics, see Swanton (2003), chapter 12.

is revealed only within a space where the diverse practices of ethics make sense, as having a unitary significance for the good life.[12]

Let us briefly illustrate the four-fold of openness by means of a virtue concept: generosity.

1. Without the general emotional attunement of benevolence, say, generosity would not be disclosed as a moral property at all. With that emotional attunement, generosity is "out in the open."
2. Without the "region" of *ethics*, a "region " that is not reducible to other regions such as that of science, and that has its own horizon of disclosure, generosity would not be understood as a *virtue*. As Hume points out, no amount of staring at a murder scene will reveal the vice in it if we do not have an ethical orientation within which such vice is intelligible as vice. Otherwise we will just see quantities of blood and so forth.
3. Without the "individual being open to the thing" by means of practical wisdom, say, generosity as a virtue could not intelligently and sensitively be applied to the world in practical decision making and comportment. Having an emotional attunement of benevolence is not enough for *expertise* in the region of ethics.
4. Without the social investing of moral properties with significance in a public, historically and culturally situated language and understandings, successful social engagement with the concept of generosity would not be possible. We would not understand how generosity as a virtue is contoured by such factors as historical and cultural situatedness, and one's role. That deployment presupposes that "humans are open to their fellows."

Finally some general features of the four-fold must be noted. First, on Heidegger's view, the fourfold of openness is objective. By means of the four-fold a facet of the world such as ethics is "brought forth." This is neither a "fabricating or a making" rather that facet we call ethics "lies before us" (Heidegger 1994, p. 76). This point is also emphasized in *Being and Time*. Certainly truth cannot exist without beings investing the world with significance; without the fourfold of *aleithia*. Hence "Before there was any *Dasein*, there was no truth; nor will there be after *Dasein* is no more" (1967, sect. 44, p. 269) Yet this is not idealism: for as Heidegger puts it, "once entities have been uncovered [by forms of disclosure proper to *Dasein*] they show themselves precisely as entities which beforehand they already were" (1967, sect. 44, p. 269). It is not that before such uncovering the entities with their disclosed properties did not exist; rather truth (as uncoveredness) was "wrested" from their hiddenness (by investing with significance).

[12] Note however that this aspect of openness has an inherent danger: a tendency to retreat from an authentic relation to the world as an individual, in which one's engagement is *one's own* (*eigentlich*), into an immersion into the average everydayness of *Das Man*. For authentic engagement, one's concern must be "anxious" to a degree. On the other hand, authentic (*eigentlich*) engagement cannot ignore this fourth aspect of openness; "openness to one's fellows." As Michael Lewis (2005) explains, authenticity and inauthenticity for Heidegger should be understood as vectors at the extremes of which are complete undifferentiation; a total lack of individuality, and extreme "existentialist" rebellion. Correct openness to one's fellows avoids these extremes.

But, nor is the world accessible by a kind of representation that is independent of that four-fold. There is no "Mirror of Nature"; certainly not for ethics. As noted above, this bringing forth is a necessary condition for the operability of truth as a property of judgments, where ethical properties are represented:

> This multiple unitary openness holds sway *in* correctness [of representation]. The openness is not first produced by the correctness of the representing, but rather just the reverse, it is taken over as what was already holding sway. Correctness of representation is only possible if it can establish itself in this openness and vaults it over. The openness is the ground and the soil and the arena of all correctness. (1994, p. 19)

Second, as the above quote intimates, although the four aspects of openness can be distinguished as above, they are unitary: "this four-fold openness would not be what it is and what it has to be if each of these opennesses were separately encapsulated from the others. This four-fold openness holds sway rather as one and unitary" (1994, p.19). We might accordingly note then, that the above separate descriptions of the fourfold of generosity as a virtue has an air of artificiality about it. The fourfold is a unity.

Application of *Aleithia* to an Environmental Virtue Ethics of Dwelling

This section applies aspects of the four-fold of opening (*aleithia*) to a virtue ethics of dwelling. Space permits consideration only of the first three, and of these only the first can be considered at some length. That will occur in the final section. Note too, that my purpose is an illustrative application to aid understanding of a Heideggerian approach in the various aspects of the fourfold of truth as *aleithia*, as applied to dwelling. Detailed critical analysis requiring confrontation with alternative approaches is not possible here.

Each aspect of the four-fold will suggest basic emotional orientations to the world and associated clusters of "dwelling virtues" some of which will be specified here. The thing being out in the open requires for Heidegger, wonder, for him a fundamental *Grundstimmung* of dwelling. Heidegger's account of this is the topic of "Heidegger's Notion of Wonder."

(1) *The thing being out in the open.*

The first aspect of the four-fold of opening is "the thing being out in the open." For Heidegger, without certain kinds of emotional disclosure, what he calls *Grundstimmungen* or fundamental emotional attunements, whole aspects of the world would remain hidden from us. Certain types of properties of objects and certain kinds of relationships would not be "out in the open." Is there an emotional attunement that reveals the world in its fundamental aspect as a place in which we dwell, a place in which we are at home? Hume's benevolence and capacity for extensive sympathy is not enough. That opens up beings as individual creatures who are sentient, who suffer, who have projects and their own lives as something to respect. Dwelling as a mode of being in the world requires something more. For Heidegger to dwell is to experience one's dwelling place as "holy." This might seem deeply mysterious to the secular mind, since holiness has religious connotations. To experience something as "holy" for Heidegger, however, is to see it as mysterious, radiant, awesome, and not something to be totally "ordered about," calculated, manipulated for our own ends. The notion of holiness connotes both a love as a desire for unity or as Kant would put it, a coming close, and what Kant would call a "keeping distance" of respect, suggested by mystery, awe, and reverence.[13] Dwelling love is a deeply comforting coming close, allowing for a "Nearness of Being" as Heidegger puts it, suggesting that one's dwelling place is a secure haven from stress, hostility, strangeness, and alienation.

The *general* notion of seeing as holy applies to all aspects of the fourfold of *aleithia*, and at the most "primordial" level for Heidegger consists of wonder, essential for the "thing being out in the open" in a dwelling orientation. A sense as holy also manifests in for example care, sense of heritage, and being at peace.

(2) *The thing is out in the open in a region between thing and man.*

The second aspect of the four-fold is "the region between thing and man."

Recall that this means that disclosures are "regional," in the sense of relative to "horizons" of disclosure. The region between thing and man signifies that the "thing," e.g., a virtue such as generosity, or compassion, or love of nature, is opened up only if it is opened within a "region" of intelligibility such as ethics, and that this region is relative to "man" in the sense that a mode of disclosure is a form of intelligibility available to and proper to human beings.

The question arises: what is the region between thing and man for dwelling? For Heidegger as I understand him, the region is simply this. Dwelling is regional in another sense. It is *local* both with respect to place and time. (Recall however that there is a universal aspect of dwelling for Heidegger.) The dimension of time for Heidegger is a retrojection back to the past, a projection into the future and a throwness into the present. Of great importance to dwelling is the regionality of the past; one's heritage. To have identity as a social being is to have a heritage. Alasdair MacIntyre is someone in the virtue ethics tradition who recognizes this, claiming:

> I am born with a past; and to try to cut myself off from that past, in the individualist mode, is to deform my present relationships...Notice that rebellion against my identity is always one possible way of expressing it. (p. 205)

The local regionality of dwelling impacts on the nature of the virtues of dwelling. In particular, a central dwelling virtue associated with the second aspect of the four-fold, is passionate devotion to regional icons reflecting both sensitivity to heritage and to place. Associated vices of excess and deficiency are respectively indifference to such icons, and fanaticism, though where the line is to be drawn between fanaticism and passionate devotion is not sharp, and a matter of controversy. Benevolence, caring, love, as *dwelling* virtues, then, are shaped by the regional nature of the orientation: "seeing as holy."

This aspect of dwelling virtue shows how (with respect to the second aspect of the four-fold of truth as *aleithia*) the horizon of disclosure of dwelling is a *logos* of

[13] For more on Kant's notions of love and respect see Swanton (2003).

the *holy*, a *logos* that is not generally conceived as present in the disclosure of ethics as such. Given that the four-fold is a unity, we need to see how the holy impacts on the regionality of space and time. This regionality suggests that what is holy has an intensely personal quality, (though a personal quality that is not disjoint from that of one's dwelling fellows (aspect (4) of the four-fold)). As a result, my examples will concern my native land, New Zealand (particularly the northern region). For example, in New Zealand, we may treat native "bush" (forest) as holy. Trees *as such* are not equal. Wilding *pinus radiata* is seen by lovers of native bush and native plants as a scourge infesting our coasts and hillsides, displacing native pohutukawa whose red flowers at Christmas time make it a very special tree, and native tussock in South Island tussock country. They are to be killed where appropriate.[14] Even areas planted on hillsides for commercial purposes alongside native bush are seen as ruining the aesthetics of our beloved bush-clad hills. In California, however, *pinus radiata* would be, and should be treated completely differently. Again a passionate conservationist of native New Zealand bush and birds may kill 100's of (Australian originated) opposums on her own land, whereas a sentience-based ethics of benevolence not so sensitive to the regional nature of "seeing as holy" might not sanction such killings. Benevolence, caring, love, as *dwelling* virtues, then, are shaped by the regional nature of the orientation: "seeing as holy."

It is also important to consider the regionality of time and heritage, as well as place. Even a passionate lover of native bush will feel outrage at any suggestion of killing the old (by New Zealand standards) English oaks growing amongst the native bush in the Auckland Domain, for these are an Auckland icon planted by the city fathers.

(3) *Being open to the thing*

Being open to the thing implies not merely that there is an attunement (a *Grundstimmung*, such as a benevolent mind-set, a sense of holiness) that as Heidegger puts it, "lays it open." It implies that we are also open to it. Like many others, I shall call this a *sensitivity* to dwelling, a sensitivity that, as one might say, constitutes a practical wisdom in relation to the holy. This is not to deny that the sense of the holy is a passionate sense, indeed once passion goes, so does "radiance." A loss of passion is tantamount to a "loss of the gods." It is not enough, therefore, to have a sense of the holy, one needs to cultivate a range of virtues that hone this sense. In general these virtues make one alert to the destruction or potential for destruction or attenuation of the holiness of one's place and its iconic symbols, objects, and practices; immune to contributing to that destruction oneself; and creative about preserving that holiness against the various dangers that it faces.

In keeping with the point made above (note 7) that dwelling orientation is a holistic orientation where the "natural" the cultural and the iconic are in relation, I shall illustrate with coach Graham Henry's justification for not being interviewed by

[14] What counts as appropriate is highly contextual, and quite properly beset by ambivalence and indeterminacy. Two huge wilding pines on our semi-rural property received a stay of execution, and I have grown to appreciate them for a variety of reasons. Now they are (almost) loved. Note that this contextualized Heideggerian position is not at all the same as treating ecosystems rather than individual organisms or species as loci of "intrinsic value."

a certain rather negative sports journalist for a book; the worry that the "mystique" of the All Blacks (the iconic New Zealand rugby team, very important to New Zealand heritage), might be undermined. If one loses sensitivity to that sort of worry—if for example one is mystified by the claim, or laughs, perhaps cynically, one is on the royal road to disenchantment. Cynicism and negativity (a vice opposed not to criticism, but constituted by selective and exaggerated negativity); readiness to dispense with "sacred" traditions, are vices opposed to dwelling virtue.[15]

Not only must this sensitivity be present, thought must also be engaged to understand the point of iconic objects and practices. For example, there have been frequently voiced claims in New Zealand that rugby is "just entertainment" competing with other forms such as going to the pictures or underwater hockey. Again, there are always calls for sports egalitarianism—related to air time and so forth, but if a sport and sports team is a "holy" icon, there is and should not be sports egalitarianism, any more than there should be "tree egalitarianism." From the perspective of Heideggerian dwelling, such dissipation can be tantamount to the "loss of the gods." One may argue that despite powerful heritage even a morally innocent icon might legitimately be changed. But the "crisis of modernity" is not that icons are changed in favor of others, but that change is too rapid, or too superficial; there is loss of the "gods" in general. The loss of a heritage is not the loss of a single object (such as an All Black who chases the money by contracting with a French Rugby club).

We may summarize the foregoing by noting that for Heidegger one might identify several layers in our understanding of "four-folds" of dwelling, each of which must be understood if virtues are to be shaped according to an ethics or orientation of dwelling. These are the following:

(a) The metaphysical foundation of dwelling: the four-fold of *truth* as *aleithia*. An understanding of this notion of truth can provide a radical critique of traditional metaphysics of ethics in the analytic tradition.
(b) The basic aspects of Dasein's *dwelling orientation* as a type of understanding and *Grundstimmung* (fundamental emotional orientation) proper to Dasein (namely seeing as holy), in both its universal and local aspects.
(c) The more detailed application of the four-fold of truth as *aleithia* to that dwelling orientation, and its virtues. This application results in four aspects of *Dasein's* dwelling-orientation as holy, and as regional relative to the dweller. A thorough and detailed application would provide us with an understanding of four sets of virtues specific to a dwelling orientation, some of which have been noted or briefly discussed here.
(d) The four-fold of the *world as dwelling place*: earth, sky, mortals, gods.

[15] Recently in New Zealand, proposals to change the rules for the challenge rugby trophy, the Ranfurly Shield, were greeted with howls of protests, despite theoretically good reasons for the change. The words "sacred" and "sacred traditions" were frequently to be heard. The New Zealand Rugby Union backed down.

Heidegger's Notion of Wonder

Wonder is one of the forms, perhaps the prime form, of emotional disclosure constituting the "laying open" of things as holy. It is a prime feature of aspect (1) of the four-fold of *aleithia*. In this "A Virtue Ethics of Dwelling" outline Heidegger's view of the nature of this orientation.

It can be understood in the first instance as the antithesis of a mind-set that Heidegger calls Enframing (*Das Gestell*). Let us describe this mindset before moving to the positive features of wonder.

The basic feature of Enframing is that it removes the mysteriousness of the world. Seeing the world as something whose mysteriousness can never be fully captured and removed by what Heidegger calls a theory of the real that renders everything totally understandable and thereby "usual," is possibly the cardinal feature of receptivity to the world as "holy." How does Enframing destroy this openness? First, what in more detail is Enframing? Enframing is a disclosure or revealing of the world exclusively as resource. It is not only a totalizing but also an absolutizing mind-set in which the world is calculated, ordered, for this purpose *alone*. The kind of revealing that is Enframing is a "challenging": that is, for Heidegger, objects are "unlocked," "transformed," "stored," "distributed," and "switched about." (Heidegger 1997, p. 16) They cannot be left alone, let be. There is nothing wrong with things being shown up as resources for use as such. For Heidegger, we are working, producing creatures for whom things are "ready-to-hand" and not just "present-at-hand." The problem occurs when the disclosure of things as resource becomes an absolutizing attunement, closing off other ways of seeing the world. Not only are objects "veiled"—not seen in all their richness—but the human being himself "exalts himself to the posture of lord of the earth." (1997, p. 27) In this way, Heidegger says, "the impression comes to prevail that everything man encounters exists only as his construct" (ibid.). Even worse, perhaps, human beings are also Enframed. They too are subjects of calculation, are assessed exclusively in terms of their quantitatively calculated outputs, and are described as "human resources" (1997, p. 18).

So what then is wonder for Heidegger? Wonder as a totalizing *Grundstimmung* (that is a *Grundstimmung* that discloses the world as a whole) at the core of environmental virtue is the complete antithesis of seeing the world exclusively as resource. It is a disposition that sees things as things in the sense described above. This is not to see things as mere things: they are seen as having significance, whether as cultural objects, or as items of nature, wondrous in their own right. Crucial to the idea of wonder is that we are never able to bring everything into the full light of understanding—the possibilities for further facets to be revealed remain open. In short, to have the idea that, for example, a scientific theory such as string theory has said everything there is to say about the world (because for example we have in our grasp the Grand Unified Theory of Everything) is to stop wondering. And to stop wondering is to stop being properly receptive to the wondrousness of the world.[16]

[16] I am not here suggesting that Heidegger's anti-science view is a true reflection on science properly conducted and understood. I do not think that science is essentially enframing, as Heidegger appears to think, but assessment of these issues is outside the scope of this paper.

For Heidegger, wondering as a fundamental attunement to the world, is not merely to be understood as inner. Such an understanding only serves to "disenchant and dispossess" (Heidegger 1994, p. 141). Indeed, given the centrality of the fourth aspect of truth as *aleithia*, the "hermeneutic turn," this would be impossible. But to deny that wonder is purely inner is not to deny that it is an emotional state of attunement. What is this emotional state? As Heidegger puts it, "the wondrous is first of all what is striking, remarkable, an exception to the habitual. We call it the curious or the amazing" (1994, p. 141). If that is the nature of wonder we may ask, how can it be a "totalizing" attunement, an attunement to the world as a whole? For is not wonder directed only at the rare and the spectacular? The understanding of wonder as totalizing is secured by contrasting wonder with several other states with which wonder might easily be confused: admiration, marveling, astonishment, awe, amazement, and curiosity. Although wonder may begin with a sense of the curious or amazing, it does not rest there, and so is not amazement proper. Amazement proper, like admiration, but unlike wonder, is not directed at the world at large but rather at specific items within it: those that are in some way or other, unusual. "To be amazed is to be carried away by something particular and unusual…" (p. 142). Admiration, too, is not totalizing. Like the objects of amazement, an object of admiration is also unusual, but it is not necessarily something that "captures curiosity, or surprise, or which enthrals or amazes" (ibid.). On the contrary, the admired is something evaluated: "the unusual that provokes admiration, the admired, becomes objective specifically explicitly *as* the unusual" (ibid.). Wonder is also to be contrasted with curiosity. Indeed in *Being and Time* curiosity is seen as an emotional orientation at odds with dwelling:

> [Curiosity] does not seek the leisure of tarrying observantly, but rather seeks the restlessness and the excitement of continual novelty and changing encounters. In not tarrying curiosity is concerned with the constant possibility of distraction. Curiosity has nothing to do with observing entities and marvelling at them… (section 36, cited in Malpas 2006, p. 75)

Finally, the objects of astonishment and awe are seen as extraordinary, in a sense that involves "a decisive suspension of position taking" (unlike admiration) (Heidegger 1994, p. 143). We do not know what to make of such objects. Wonder, as a disposition concerning the world at large, cannot be identified with those states, though it may begin with them.

How then is wonder to be distinguished from these dispositions, as a totalizing, fundamental, attunement: a *Grundstimmung*? As Heidegger claims:

> Insofar as this disposition turns to the whole and stands in the whole, it is called a basic disposition. (1994, p. 145)

Wonder concerns the world at large because it concerns the usual, but the usual seen as unusual. Hence it *begins* with a sense of the striking, the remarkable, the amazing. In wonder, the usual arises as unusual, and as such, provokes questioning. In a state of wonder, "we find ourselves in the midst of beings as such and as a whole" and we find ourselves "caught up in them" (p. 147). Not only then, is wonder a fundamental disposition concerning the world as a whole, and one that provokes basic questioning, it is a disposition that should not eventually cease to be totalizing in this manner. For in a state of wonder, one does not think that, as a terminus or goal of wonder, the usual should revert to being usual, having become unusual. As Heidegger puts it, we know "no way out" of the unusualness of what has become most unusual (p. 144). That is, we should not imagine that a world view such as THE theory of the real will finally allow us to see and understand all there is to be understood, and make things quite usual again. Rather, "as moved by wonder, man must… experience and sustain … unconcealedness as the primordial essence of beings" (p. 146). The very fact that the questioning is to be *sustained*, shows that it does not terminate in a *complete* revealing of the world, and also that, if it ceases, the world will lapse into a kind of darkness (concealment). Wonder guards against both an absolutizing view of the world in which all is thought to be revealed and made usual, and against a kind of Dark Age.

Wonder then, is an antidote to Enframing. As Young (2002) points out, for Heidegger, it is of the essence of technology that it poses a threat in the form of a potential transformation into the absolutizing disclosure of world as resource, for the everyday work-oriented disclosure of the world in relation to us may become familiar, excessively so. Unless we are taken out of this state in wonder, this familiarity can become absolutized. That is, the world cannot be seen in any way other than as work-related resource. However, as we have seen, in wonder, the usual becomes the unusual, and that in turn drives out the tendency of the familiar to be swallowed up into the totalizing attunement of Enframing. The world can remain mysterious. An attunement that leaves one open to the mysteriousness of the world is destroyed by "dry," "monotonous," and "therefore oppressive" resource thinking. (Heidegger 1997, p. 17).[17]

References

Dreyfus, H. L. (2006). Nihilism, art, technology, and politics. In C. Guignon (Ed.), *The Cambridge companion to Heidegger* (2nd ed., pp. 345–372). Cambridge: Cambridge University Press.

Follett, K. (2007). *World without end*. London: MacMillan.

Frasz, G. (2005). Benevolence as an environmental virtue. In R. Sandler & P. Cafaro (Eds.), *Environmental virtue ethics* (pp. 121–134). Lanham: Rowman and Littlefield.

Guignon, C. (Ed.). (2006). *The Cambridge companion to Heidegger* (2nd ed.) Cambridge: Cambridge University Press.

Heidegger, M. (1967). *Being and time* (J. Macquarrie & E. Robinson, Trans.). Oxford: Basil Blackwell.

Heidegger, M. (1971a). Poetically man dwells. In M. Heidegger (Ed.) *Poetry, language, thought* (A. Hofstadter, Trans.). (pp. 213–229). New York: Harper and Row

Heidegger, M. (1971b). Building dwelling thinking. In *Poetry, language, thought* (pp. 145–161).

Heidegger, M. (1971c). The thing. In *Poetry, language, thought* (pp. 165–186).

Heidegger, M. (1975). *Discourse on thinking* (J. Anderson & E. Freund, Trans.). New York: Harper and Row.

Heidegger, M. (1994). *Basic questions of philosophy: Selected problems of logic* (R. Rojcewicz & A. Schuwer, Trans.). Bloomington: Indiana University Press.

[17] I thank Philip Cafaro for helpful suggestions, and Julian Young who, through his writings and conversation, has helped greatly in opening my mind to the riches Heidegger has to offer.

Heidegger, M. (1995). *The fundamental concepts of metaphysics: World, finitude, solitude* (W. McNeill & N. Walker, Trans.). Bloomington: Indiana University Press.
Heidegger, M. (1997). The question concerning technology. In *The question concerning technology and other essays*, (W. Lovett, Trans.). (pp. 3–35). New York: Garland Publishing.
Hoy, D. C. (2006). Heidegger and the hermeneutic turn. In C. Guignon (Ed.), *The Cambridge companion to Heidegger*, (pp. 177–201).
Lewis, M. (2005). *Heidegger and the place of ethics*. London: Continuum.
MacIntyre, A. (1981). *After virtue: A study in moral theory*. London: Duckworth.
Malpas, J. (2006). *Heidegger's topology: Being, place, world*. Cambridge: MIT Press.
McDowell, J. (1995). Two sorts of naturalism. In R. Hursthouse, G. Lawrence, & W. Quinn (Eds.), *Virtues and reasons* (pp. 147–179). Oxford: Clarendon Press.
Rolston, H. (2005). Environmental virtue ethics: Half the truth but dangerous as a whole. In R. Sandler & P. Cafaro (Eds.), *Environmental virtue ethics*, (pp. 60–78).
Sandler, R. (2007). *Character and environment: A virtue oriented approach to environmental ethics*. New York: Columbia University Press.
Sandler, R., & Cafaro, P. (Eds.). (2005). *Environmental virtue ethics*. Lanham: Rowman and Littlefield.
Scheffler, S. (1992). *Human morality*. New York: Oxford University Press.
Schmidtz, D. (1998). Are all species created equal? *Journal of Applied Philosophy, 15*, 57–67.
Swanton, C. (2003). *Virtue ethics: A pluralistic view*. Oxford: Oxford University Press.
Taylor, P. (1986). *Respect for nature: A theory of environmental ethics*. Princeton: Princeton University Press.
Taylor, C. (2006). "Engaged agency and background in Heidegger," in C. Guignon (ed.), *The Cambridge companion to Heidegger*, (pp. 202–221).
Timmons, M. (1999). *Morality without foundations: A defense of ethical contextualism*. Oxford: Oxford University Press.
Weberman, D. (1996). Heidegger and the disclosive character of the emotions. *The Southern Journal of Philosophy, 34*, 379–410.
Wrathall, M. A. (2006). "Truth and essence of truth in Heidegger's thought," in C. Guignon (ed.), *The Cambridge companion to Heidegger*, (pp. 241–267).
Young, J. (1999). Artwork and sportwork: Heideggerian reflections. *Journal of Aesthetics and Art Criticism, 57*, 267–277.
Young, J. (2002). *Heidegger's later philosophy*. Cambridge: Cambridge University Press.
Young, J. (2006). "The fourfold" in C. Guignon (ed.), *The Cambridge companion to Heidegger*, (pp. 373–392).

J Agric Environ Ethics (2010) 23:167–183
DOI 10.1007/s10806-009-9203-4

ARTICLES

Ethical Theory and the Problem of Inconsequentialism: Why Environmental Ethicists Should be Virtue-Oriented Ethicists

Ronald Sandler

Accepted: 27 July 2009 / Published online: 7 August 2009

Abstract Many environmental problems are longitudinal collective action problems. They arise from the cumulative unintended effects of a vast amount of seemingly insignificant decisions and actions by individuals who are unknown to each other and distant from each other. Such problems are likely to be effectively addressed only by an enormous number of individuals each making a nearly insignificant contribution to resolving them. However, when a person's making such a contribution appears to require sacrifice or costs, the problem of inconsequentialism arises: given that a person's contribution, although needed (albeit not necessary), is nearly inconsequential to addressing the problem and may require some cost from the standpoint of the person's own life, why should the person make the effort, particularly when it is uncertain (or even unlikely) whether others will do so? In this article I argue that justifications for making the effort to respond to longitudinal collective action environmental problems are, on the whole, particularly well supported by virtue-oriented normative theories, on which character traits are evaluated as virtues and vices consequentially or teleologically and actions are evaluated in terms of virtues and vices. If ethical theories are to be assessed on their theoretical and practical adequacy, and if providing a compelling response to the problem of inconsequentialism is an instance of such adequacy, then this is a reason for preferring virtue-oriented ethical theory over non-virtue-oriented ethical theories, such as Kantian, act utilitarian, and global utilitarian theories.

Keywords Virtue-oriented ethics · Utilitarianism · Kantian ethics · Global environmental problems

R. Sandler (✉)
Department of Philosophy and Religion, Northeastern University, 371 Holmes Hall, Boston, MA 02115-5000, USA
e-mail: r.sandler@neu.edu

Introduction

Many significant environmental problems are *longitudinal collective action environmental problems*. They arise from the cumulative unintended (and often unforeseen or unforeseeable) effects of a vast amount of seemingly insignificant decisions and actions by individuals who are unknown to each other and distant from each other (spatially, temporally, and socially).[1] Global climate change is perhaps the paradigmatic longitudinal collective action environmental problem. However, many other environmental issues, such as pollution, fisheries depletion, human population growth, deforestation, and ozone layer degradation, exhibit these features as well. It takes quite a lot of people, almost all of them acting without environmental malice (though often with environmental ignorance or indifference), to precipitate the environmental problem.[2]

Longitudinal collective action environmental problems are likely to be effectively addressed only by an enormous number of individuals each making a nearly insignificant contribution to resolving them. However, when a person's making such a contribution appears to require social, personal, or economic costs the *problem of inconsequentialism* arises: given that a person's contribution, although needed (albeit not necessary), is nearly inconsequential to addressing the problem and may require some cost from the standpoint of the person's own life, why should the person make the effort, particularly when it is uncertain (or even unlikely) whether others will do so?

A common response to the problem of inconsequentialism is to emphasize the need for political solutions to longitudinal collective action problems and, therefore, the need for political activism on the part of individual agents, i.e., acting in support of developing adequate political arrangements and sufficient compliance incentives to address the problems. However, even if it is true that longitudinal collective action environmental problems are likely to be resolved only through political arrangements and that acting in support of developing these is justified, this does not adequately address the problem of inconsequentialism. Consider, for example, global climate change. The problem of inconsequentialism arises regarding activism in support of political arrangements and institutions to address global climate change, since the activities of the vast majority of us are inconsequential to bringing about the national and international agreements and activities necessary to address the problem. Moreover, that one ought to actively support aggressive national greenhouse gas emission reduction policies (or other systematic changes related to

[1] Not all environmental problems are longitudinal collective action problems. Some widespread environmental problems were brought about by the actions of a relative small number of people, e.g., problematic invasive species such as milfoil or gypsy moth. Other environmental problems or issues are localized, e.g., neighborhood green space management.

[2] The cumulative or collective aspect of these problems is due to their being tied to consumption of resources. Any one of us—even the very wealthy of us—can consume only a very small, seemingly inconsequential amount of the overall planetary or regional resources; and, with few exceptions, each person has very little control over the consumption patterns of others. The distance aspects of these problems are mediated by climactic and ecological interconnectedness. The problems are longitudinal because alterations of climactic and ecological systems can take considerable time to mature and spread, but often are persistent once they do.

fossil fuel consumption) does not tell a person what she ought to do with respect to reducing her own emissions, particularly in the absence of such policies. In addition, even if there is a political arrangement in place, the problem of inconsequentialism still arises if the costs of non-compliance are lower than the costs of compliance (or the benefits are greater than those of compliance). Thus, the problem of inconsequentialism arises when there is no political arrangement in place to address a longitudinal collective action environmental problem, there is such an arrangement but it is inadequate to address the problem, or there is an adequate arrangement but the costs to the agent of contributing are greater than those of not complying (even given the arrangement's mandates, sanctions, and incentives to encourage participation); and appeals to the need for political activism (or promoting systematic change) do not address what, if anything, one ought to do with respect to the problem itself.

The problem of inconsequentialism is old news in environmental ethics, and several types of responses to it have been offered. *Appeal to agent benefit* responses attempt to show that making an effort to address longitudinal collective action environmental problems are not as costly to the agent as might initially appear, e.g., that making the effort to address the problem is itself beneficial to the person or that, while there are some initial costs, making the effort to address the problem is beneficial to the person in the long run.

Appeal to duty responses attempt to shift the focus of assessment from the costs and benefits involved in acting to the acting itself, e.g., that a person is required to do their duty with respect to the problem, regardless of whether others do theirs; that a person is required to act in ways that would resolve the problem if others were to act likewise; or that a person is required to do their fair share to address the problem, no matter how small the share.

Appeal to amplifying effects responses attempt to show that an individual's contributions to addressing longitudinal collective action problems are not as inconsequential as they might initially appear, e.g., that in making the effort, a person sets an example that could lead to more widespread participation to address the problem; that the sum of a person's actions over her lifetime is not inconsequential to the problem; that if systematic change to address the problem is to occur, it must be preceded, precipitated, and lead by committed individuals.

Appeal to character responses attempt to shift the focus of assessment from the costs and benefits involved in acting to the agent herself, e.g., that making the effort to address the problem is a matter of personal integrity; that to fail to make the effort to address the problem is to be complicit with those who do not; or that to fail to make the effort to address the problem is to be indifferent or insensitive to the problem.

In this article I argue that this array of justifications for making the effort to respond to longitudinal collective action environmental problems is, on the whole, particularly well supported by virtue-oriented normative theories, on which character traits are evaluated as virtues and vices consequentially or teleologically and actions are evaluated in terms of virtues and vices. The broader theoretical context or meta-normative underpinnings of such a normative theory could be virtue ethical (e.g., neo-Aristotelian) or utilitarian (e.g., virtue utilitarian). If ethical theories are to be assessed on their theoretical and practical adequacy, and if providing a compelling response to the problem of inconsequentialism is an instance of such adequacy, then this is a reason for preferring virtue-oriented ethical theory over non-virtue-oriented Kantian and utilitarian theories (e.g., act utilitarianism and global utilitarianism).

Here, then, is a summary of this article's central argument:

(1) Many significant environmental problems are longitudinal collective action problems that give rise to the problem of inconsequentialism.
(2) An adequate environmental ethic will justify acting to address significant environmental problems.
(3) Therefore, an adequate environmental ethic will provide a cogent response to the problem of inconsequentialism—i.e., it will explain why a person ought to make the effort (or take on the costs) to make a (even insignificant) contribution to addressing longitudinal collective action environmental problems.
(4) Virtue-oriented ethical theory supports well several types of responses to the problem of inconsequentialism, whereas non-virtue-oriented forms of utilitarianism and Kantianism do not.
(5) Therefore, virtue-oriented ethical theory is preferable to those theories (and environmental ethicists should be virtue-oriented ethicists), all other things being equal.

Thus far I have attempted to motivate 1–3. The remainder of the article will focus on premise 4 and the inference to 5. I begin by identifying the shortcomings of act utilitarian and Kantian ethics with respect to longitudinal collective action problems, and use these to help articulate what the structure of a normative ethic well resourced for addressing the problem of inconsequentialism would be like. I then argue that virtue-oriented ethical theory exhibits that structure. Finally, I discuss two utilitarian alternatives to virtue-oriented ethical theory, emulator utilitarianism, and global utilitarianism.

Act Utilitarianism

According to act utilitarianism, an action is right to the extent that it brings about the best (or good enough) consequences of all the courses of action available to the agent.[3] Act utilitarian reasoning provides the clearest formulation of the problem of inconsequentialism. Almost any action performed by almost any agent will have a

[3] Varieties of act utilitarianism are generated by different value axiologies and different accounts of best (or good enough) consequences (e.g., proportional, absolute, or minimizing bad). For example, a hedonistic value axiology yields the following version of act utilitarianism: "actions are right in proportion as they tend to promote happiness, wrong as they tend to produce the reverse of happiness. By happiness is intended pleasure and the absence of pain; by unhappiness, pain, and the privation of pleasure" (Mill 1972, p. 7). In what follows, a generic version of act utilitarianism is considered, one that applies the principle of utility at the level of individuals acts, without a specified value axiology or account of best (or good enough) consequences.

vanishingly small effect on longitudinal collective action environmental problems.[4] Many of these same actions will have burdens for the agent and those close to her (i.e., family and friends), in terms of, for example, time, economic costs, social costs, and professional costs. In such cases the local utility of actions that contribute to longitudinal collective action problems (or that fail to address the problems) will outweigh the inconsequential global utility of those actions. Therefore, act utilitarianism cannot explain why we ought to act or live in ways responsive to longitudinal collective action problems when doing so has costs or sacrifices associated with it. This suggests two features that a normative theory well resourced for addressing the problem of inconsequentialism will exhibit:

(1) That discrete actions are not evaluated entirely on the basis of the outcomes (or likely outcomes) of the action.
(2) That it evaluate on the basis of patterns of behavior or activities throughout a person's life or patterns among people or communities.

Advocates of act utilitarianism might respond to the claim that act utilitarian reasoning gives rise to the problem of inconsequentialism by arguing that it trades on a mistaken version of the theory. It is not the consequences forecasted by agents that are relevant to the rightness of actions, but the action's actual consequences (or, perhaps, what they are in fact likely to be). Therefore, it is open to act utilitarians to argue that while many agents might believe that their actions to address longitudinal collective action environmental problems have high local costs (and low local and global benefits) often this is not the case. That is, they might offer versions of appeal to agent benefit or amplifying effects responses. However, both these types of responses have significant limitations in the context of act utilitarian theory. For appeal to agent benefits responses, the key limit is that the local (dis)utilities associated with the effort made by an individual to address some environmental problem are likely to be related to her particular perspectives. A person who does not care much about the environmental problem is not likely to feel good (a local utility) about having made an effort to address the problem. A materialistic person who lives a high consumption lifestyle is more likely to feel constrained (a local disutility) by lower levels of consumption. Act utilitarianism has difficulty overcoming this limitation because it does not evaluate the appropriateness of an agent's perspectives or have such evaluations inform evaluations of the rightness of action. This suggests that a normative theory resourced for supporting appeal to agent benefit responses to the problem of inconsequentialism will have the following features:

(3) That its evaluative focal points (i.e., what it directly evaluates) include attitudes or perspectives of people.
(4) That evaluation of attitudes and perspectives bear on evaluation of actions.

[4] Exceptional cases include those with significant social or political authority or influence over others (e.g., President of the United States or the Catholic Pope), as well as those who create and disseminate significantly powerful relevant technologies, for example.

Appeal to amplifying effects responses, e.g., that by moving others to act or setting a habit/trajectory in one's own life a person's actions are not really inconsequential to longitudinal collective action environmental problems—are also limited within act utilitarianism. With respect to effects on others, it often will not be the case that what a single person does in a single act will impact the actions of others or be crucial to bringing about systematic reform, particularly in situations where others are not already acting responsively or motivated to do so. With respect to the cumulative effects over a person's lifetime, even a person's entire lifetime impacts on any longitudinal collective action problem is likely to still be inconsequential. Moreover, refocusing evaluation from individual acts to lifetimes requires moving to a non-act version of utilitarianism.

In addition, and more important for present purposes, act utilitarian appeals to amplifying effects responses fail to explain why one should make the effort to address the problem even if others are not (or are not likely) to do so. Act utilitarianism, since it evaluates discreet actions indexed to a particular time, place, and context, is not able to decouple the rightness of action from what others are doing (or are likely to do). If act evaluation is not indexed to the actual world or possible worlds relevantly similar to the actual world, act utilitarianism sacrifices its intuitive plausibility, which rests on the idea that the right action is the one that would bring about the best consequences of those available to the agent at this time and place. It is for this reason, i.e., that rightness of action is tied to what others do or are likely to do, that act utilitarianism does not support well several of the duty-oriented responses to the problem of inconsequentialism, e.g., that one should do one's fair share even if others do not do theirs. On act utilitarianism, what one agent ought to do often heavily depends on what other agents are doing (or are likely to do). This limitation suggests an additional feature that a normative theory well resourced for addressing the problem of inconsequentialism will have:

(5) That evaluation of an agent's actions is not overly contingent on the actions of others.

There are three primary ways that an advocate of act utilitarianism might respond to the set of criticisms presented above. First, she might argue that it is not a criterion of adequacy for an environmental ethic that it provides normative resources to respond effectively to the problem of inconsequentialism.[5] Second, she might argue that some or all of the features identified in this section as conducive to an ethical theory's capacity to respond to the problem of inconsequentialism are not needed. Third, she might argue that some version of act utilitarianism can exhibit some or all of the features. I have already addressed the first and second of these responses by providing reasons for believing it is an adequacy criterion and that the

[5] Baylor Johnson has argued that taking on costs associated with individual responsiveness to a collective action environmental problem is not rational if one's actions are inconsequential and not part of a collective effort to address the problem (Johnson 2003). However, as discussed above, even if there is a collective effort in place, the problem arises: Why should one make the effort, given that one's contribution is insignificant to whether it will succeed? What drives the problem is not the absence of a collective response, but the insignificance of one's impact on the problem and the costs associated with responding to it.

features identified are crucial to meeting it. With respect to the third response, it may be that a sufficiently sophisticated version of act utilitarianism can exhibit the features. However, such an act utilitarianism would be (in the terminology of this article) a virtue-oriented form of act utilitarianism. Therefore, it would not belie the central argument of this article, which aims to show that virtue-oriented ethical theory, regardless of the basic theoretical construct out of which it is developed, is particularly well resourced for responding to the problem of inconsequentialism. The target of the arguments in this section is non-virtue-oriented forms of act utilitarianism.

Kantian Ethics

Kantian ethics appears less susceptible to the problem of inconsequentialism, since it evaluates actions on the basis of their maxims (i.e., principles from which they are performed or that they express), rather than on the basis of what they bring about. The fact that an agent's actions might be inconsequential in resolving (or creating) an environmental problem is, therefore, not material to the evaluation of the action. It is one's duty to act on appropriate maxims, those that conform to the moral law, i.e., that can be consistently willed to be universal law and that do not treat something with inherent worth as a means only and not an end in itself, without regard to the consequences (or lack thereof) of an act (Kant 1997). In this way, Kantian ethics appears to reject the sort of reasoning that gives rise to the problem of inconsequentialism.

However, Kantian ethics struggles with a different dimension of longitudinal collective action environmental problems: that the problems often result from unintended byproducts of activities, where those byproducts are not a necessary means to the ends sought. Consider, for example, global warming. The continuing build up of greenhouse gases in the atmosphere is not sought by those whose actions contribute to it, nor is it a means to any end that is sought. Carbon dioxide emissions are not a means to producing energy; it is a byproduct of the fossil fuel energy production process. If the process did not produce carbon dioxide, it would make no difference to the point of the process or to whether its aims are achieved. Moreover, when a person contributes, through her consumption, to the amount of greenhouse gases in the atmosphere, those who are affected by the resultant climactic and environmental changes do not stand as a means to the consumer's ends, e.g., health, convenience, comfort, recreation, or status. If the emissions did not occur, did not contribute to global climate change, or did not result in ecological and social problems, the consumptive actions and practices would still have the same intended ends and means of accomplishing them.

Because longitudinal collective action environmental problems are a by-product of activities that have other intended ends and do not stand as means to those ends, Kantian ethics has difficulty explaining what is wrong with the actions or practices that give rise to them. If one buys a terribly inefficient vehicle to take the kids to soccer practice, to raise one's status, or just because one likes driving it there is no person (or animal or plant, for that matter) treated as a mere means in one's doing so. It does not violate or offend the value of other human beings (or animals or plants). There is no contradiction in willing the maxim as one appropriate for anyone (or everyone) else. The same is true of having children and population growth. The ecological challenges associated with global population increases are not a sought end or a means to any sought end of those who choose to have children.[6] Therefore, those ecological challenges do not inform the maxim (or principle) that is evaluated. This suggests a crucial feature that a normative theory well resourced to respond to the problem of inconsequentialism will exhibit:

(6) That evaluation of actions is sensitive to the significance of consequences, even when those consequences are unintended byproducts of actions.

One way that a Kantian might attempt to incorporate this feature is by focusing on the possibility of moral agency itself. That is, a Kantian might emphasize that longitudinal collective action problems can undermine necessary conditions of moral agency, e.g., material goods, such as potable water and food, needed for adequate cognitive functioning (Westra 2005). Therefore, acting on a maxim that if universally adopted gives rise to them violates the moral law, since it renders acting on maxims itself impossible or involves disregarding the value of rational agency. However, while the environmental problems associated with global warming (and other longitudinal collective action environmental problems) are likely to be severe, they will not make moral agency (particularly for those who are highly consumptive, and so well resourced to adapt) impossible. Nor will they make the behaviors that gave rise to the problem, e.g., driving fuel inefficient vehicles, impossible. So, with respect to most (if not all) actions that contribute to longitudinal collective action environmental problems, universalizing their maxims would not result in undermining moral agency or otherwise disregarding the value of rational agency.

An advocate of Kantian ethics might concede that it is difficult to establish a perfect duty to refrain from contributing to collective action environmental problems, but argue that Kantian ethics can generate at least imperfect or meritorious duties not to contribute to (and to address) them. One way this might be done is by focusing on the ways in which they rely on a limited resource base. However, if the finitude of resources needed for performing an action are grounds for not being universalizable, e.g., that a maxim to travel in whatever vehicle one pleases, including those that are inefficient and run only on fossil fuels, cannot be universalized because there is only a finite amount of fossil fuels—then Kantian ethics will be reduced to absurdity. It would mean that farming violates the moral law due to the finitude of farm land and being a practicing doctor violates the moral law due to the finitude of patients. So this reply succeeds in establishing a duty to respond to longitudinal collective action environmental problems at the cost of rendering Kantian ethics non-viable.

[6] Not all longitudinal collective action environmental problems are this way. For example, with deforestation to create farmland, the deforestation is a necessary means to the end sought; and factory farming involves treating animals as a means to a sought end. So, responsiveness to some longitudinal collective action problems can be justified (even required) by some versions of Kantian ethics (e.g., one on which animals are considered ends in themselves; Korsgaard 2004; Regan 1983).

Another way a Kantian might seek to establish an imperfect duty to respond to longitudinal collective action environmental problems is by appeal to the duty to be benevolent. On Barbara Herman's interpretation of Kantian benevolence, "What one may not do is fail to give aid when true needs are threatened *for* reasons of mere self-interest" (Herman 1993, p. 156). If not giving aid for reasons of self interest is impermissible, contributing to a threat to others for reasons of self interest would seem a fortiori impermissible. Thus, failing to reduce one's greenhouse gas emissions, and thereby contributing to the problems associated with global warming, for reasons of self-interest would appear to violate the Kantian duty of benevolence. However, this response is limited for several reasons. First, even on Herman's view, a person does not fail to act benevolently if she is not capable of providing the necessary aid. As Herman puts it, "Still one may fully acknowledge the claim of need but be in no position to help because one does not possess what is needed" (Herman 1993, p. 156). This is the situation with respect to longitudinal collective action environmental problems. For the vast majority of us, our actions are inconsequential to the problem. We do not "possess what is needed." Second, if we must only sometimes act responsively, as is the case with imperfect duties, it is not likely that longitudinal collective action environmental problems will be effectively addressed. Third, because benevolence is an imperfect duty, it could be satisfied in ways independent of longitudinal collective action environmental problems, e.g., by volunteering at soup kitchens or tutoring in need children. Therefore, an imperfect duty of benevolence (or even an imperfect duty to address longitudinal collective action environmental problems) will not require responsiveness adequate to the problem.

The type of response to the problem of inconsequentialism most at home within Kantian ethics is appeal from duty. But this response only works if one's duty (or the domain of justice or fairness) involves acting to address longitudinal collective action environmental problems. It turns out, for the reasons discussed above, that it does not.[7]

As with act utilitarianism, there are three primary ways that an advocate of Kantianism might respond to the criticisms presented above. First, she might argue that it is not a criterion of adequacy for an environmental ethic that it provides normative resources to respond effectively to the problem of inconsequentialism. Second, she might argue that having unintended byproducts inform evaluation of actions is not needed to provide responsiveness to the problem of inconsequentialism. Third, she might argue that some version of Kantian ethics can exhibit this feature. I have addressed the first two of these above. However, with respect to the third, if a version of Kantian ethics can incorporate this and the other features

[7] An additional difficulty for Kantian ethics with respect to longitudinal collective action problems is that the problems characteristically involve future generations. Future generations are non-existent, so the individuals that comprise them cannot be treated as a means to an end by anything that we do. Moreover, individuals in future generations cannot be harmed or benefitted. They can be badly off, but not worse off than they otherwise would have been, since they would not have come into existence under any other scenario. As a result, Kantian ethics have difficulty supporting consideration of future generations as ethically relevant. Thus, to the extent that longitudinal collective action environmental problems involve the welfare of future generations (whoever they turn out to be) and this is ethically significant, Kantian ethics has difficulty fully characterizing the ethical dimensions of the problems.

conducive to addressing the problem of inconsequentialism, then it would be a virtue-oriented form of Kantianism, and as such would not belie this article's central argument. The point of this and the preceding section was not to show that no normative theory developed out of act utilitarian or Kantian ethical frameworks can address the problem of inconsequentialism. It was to demonstrate why standard formulations of those types of theories fail to adequately do so, and use this as a basis for generating an account of features that a normative theory well resourced for responding to the problem would have.

Virtue-Oriented Ethics

The limitations of familiar versions of act utilitarian and Kantian ethics with respect to longitudinal collective action environmental problems suggested that a normative theory well resourced for responding to the problem of inconsequentialism would have the following features:

(1) That discrete actions are not evaluated entirely on the basis of the outcomes (or likely outcomes) of the action.
(2) That it evaluate on the basis of patterns of behavior or activities throughout a person's life, as well as patterns among people or communities.
(3) That its evaluative focal points (i.e., what it directly evaluates) include attitudes or perspectives of people.
(4) That evaluation of attitudes and perspectives bear on evaluation of actions.
(5) That evaluation of an agent's actions is not overly contingent on the actions of others.
(6) That evaluation of actions is sensitive to the significance of consequences, even when those consequences are unintended byproducts of actions.

These features characterize a two-tiered virtue-oriented normative theory, where tier one is teleological or consequential evaluation of character traits and tier two is evaluation of actions, practices, and policies in light of those traits. Here is a generic formulation of such a theory:

> *Theory of virtue*: A character trait is a virtue to the extent that its possession is generally conducive to promoting the good; and a character trait is a vice to the extent that it is generally detrimental to promoting the good.
> *Principle of right action*: An action is right to the extent that it is virtuous.

This theory of virtue and principle of right action are severely underspecified. Among the features in need of explication are: what constitutes a character trait; the level of generality at which traits are evaluated (e.g., all moral agents, all human beings, particular cultures, or particular cultures in particular environmental circumstances); the class of worlds over which traits are evaluated (e.g., the actual world, possible worlds very similar to the actual world, or worlds in which all relevant agents possess the trait); the nature and source of the good; what is constitutive of the good; what it is to promote the good; and what it is for an act to be virtuous (e.g., what a virtuous person would do, what a virtuous person would advise one to do, or that it hit the target of the operative virtues).

Nevertheless, this generic theory of virtue and principle of right action exhibit the features enumerated above. The theory of virtue evaluates attitudes or perspectives of people directly (3), and it does so on the basis of patterns of behavior throughout a person's life and among people generally (2). Actions are evaluated in terms of the virtues (4). Because the theory of virtue evaluates character traits on the basis of outcomes, promoting the good, even when those outcomes are unintended or byproducts and actions are evaluated in terms of the virtues, act evaluation is informed by unintended consequences (6), without acts being evaluated only on the basis of their consequences (1). Finally, evaluations of actions are not highly contingent on the actions of others, since an act is evaluated on the extent to which it is virtuous, which is independent of whether or not others act virtuously (5).

Because it exhibits these features, such a virtue-oriented ethical theory supports well several of the reasons proffered for being responsive to longitudinal collective action environmental problems. It clearly does with respect to appeal to character responses, since whether an action is virtuous bears directly on its rightness. If an act exhibits indifference to the welfare of others (possibly including non-humans), and indifference of this sort is a vice, then it is at least prima facie wrong to act in that way. The same holds for other character traits that advocates of appeal to character responses might invoke, such as complicity, fairness, or personal integrity. If these character traits are virtues and failing to alter one's behavior in response to longitudinal collective action environmental problems is contrary to them, then according to a virtue-oriented approach, a person acts wrongly if she does not act responsively.

Appeal to duty responses are also well supported by virtue-oriented ethical theory. Within a virtue-oriented theory, an action is right to the extent that it is virtuous and not vicious, wrong to the extent that it is vicious and not virtuous. Categories of "obligatory," "permissible," and "supererogatory" can be generated by setting thresholds along the continuum of virtuousness and viciousness.[8] There is a threshold below which actions within certain domains (e.g., those with implications for the welfare of other people) are sufficiently vicious or lacking in virtue that they are not permissible, and this generates duties or obligations not to act in those ways (Swanton 2003). Thus, virtue-oriented theory has the capacity to ground duties or obligations. Moreover, whether something is obligatory is a matter of the extent to which it is virtuous or vicious, which, as discussed above, is not contingent on whether other people are acting virtuously or viciously. So even if others are not doing their fair share (or acting in ways that would, if everyone so acted, address the problem) virtue-oriented ethical theory can support the claims that a person has an obligation or duty to do hers.

Appeal to agent benefit responses are also well supported within a virtue-oriented approach. Many virtue-oriented theories are eudaimonistic or, at least, include

[8] A virtue-oriented ethical theory need not include a continuum theory of right action on which rightness and wrongness is a matter of degree. However, among the considerations in favor of a continuum theory is that it can provide these categories (Sandler 2007). Also, some virtues and vices (e.g., those that are self-regarding) might not be as relevant to these categories as others (e.g., those that are other-regarding).

eudaimonia (or agent welfare or well-being) among a plurality of goods that virtue is conducive to promoting (Sandler 2007). Moreover, because virtue-oriented ethics evaluates perspectives and outlooks (i.e., character traits) directly, it can evaluate agent benefit counterfactually with respect to her perspectives. For example, there is a considerable body of empirical evidence that suggests that it is detrimental to a person to be highly materialistic, i.e., to value highly the possession of material goods in one's life, be desirous to possess them, be emotionally invested in their presence and absence, and orient one's life to accumulate them (Kasser 2002; Kasser and Kanner 2003). If a person is highly materialistic, it might be false that reducing consumption would be better for her, given her dispositional background; but it is probably true (given the research) that it would be better for her not to be highly materialistic. Furthermore, a person who is not materialistic, and prioritizes relationships and self-realization in her life, will not see reducing consumption as a sacrifice. This is an instance of the integrating effect of virtue. For a person who possesses the virtues, acting virtuously, even when it appears from other perspectives to involve limitation and sacrifice, is in general beneficial and regarded as such (McDowell 1980; Foot 2001).

The final type of response to the problem of inconsequentialism is the *amplifying effects response*. There are limitations to this type of response in genera, i.e., independent of their being situated within a particular ethical theory. For example, even the cumulative effects of a person's actions over her lifetime are usually inconsequential in their impact on many longitudinal collective action problems; and it is not often the case that what a single person does impacts the behavior of many others or is crucial to bringing about systematic reform. However, to the extent that certain dispositions are conducive to bringing about such amplifying effects, they will inform character trait evaluation within virtue-oriented ethical theory. Moreover, virtue-oriented ethical theory has the resources to justify modifying one's behavior in response to longitudinal collective action problems even if other are not likely to do so, which is crucial if one is to set an example or lead others to change. Therefore, the *amplifying effects responses* are reasonably well supported within a virtue-oriented framework, even as they are, in general, not as compelling or robust as the other types of responses.

Thus, overall, virtue-oriented normative theory is significantly better resourced for meeting the challenge of inconsequentialism than are standard versions of Kantian ethics and act utilitarianism. Therefore, environmental ethicists should be virtue-oriented ethicists, all other things being equal. This is not to claim that environmental ethicists should be virtue ethicists (as the term is often used). It is possible develop a virtue-oriented normative theory from neo-Aristotelian and other virtue ethics approaches to ethics. But there are also forms of utilitarianism that are virtue-oriented ethical theories. It is possible, for example, to construct a form of *virtue utilitarianism* on which a character trait is a virtue to the extent that it is conducive to promoting utility and an action is right if it is virtuous (e.g., done from or expresses virtue). Similarly, it is possible to construct a version of *rule-utilitarianism* that is a virtue-oriented normative theory. Brad Hooker, for example, has defended a version of rule internalization utilitarianism (he calls it rule consequentialism) on which an action is right if it is consistent with a code of rules the general internalization of which promotes consequences as good or better than any alternative code (Hooker 2000). If internalization is understood dispositionally (i.e., in terms of character traits), this is a virtue-oriented normative theory. As discussed above, it may even be possible to formulate virtue-oriented forms of Kantian ethics and act utilitarianism.

Emulator Utilitarianism, Global Utilitarianism, and Virtue-Oriented Ethical Theory

Dale Jamieson has recently advocated a version of utilitarianism that might be called *emulator utilitarianism*. Emulator utilitarianism is, like all utilitarianisms, committed to bringing about what is best [or, at least, what is good enough or "more good than not" (Driver 2001, p. 91)]. However, it is not committed to any particular theoretical framework or level of evaluative focus. On Jamieson's view, "Utilitarianism is a universal emulator: it implies that we should lie, cheat, steal, even appropriate Aristotle, when that is what brings about the best outcomes" (Jamieson 2009, p. 1). Jamieson argues, as I have, that Kantian ethics is not conducive to justifying appropriate responsiveness to global environmental challenges. He further argues, as I have, that the nature of global environmental challenges favors an evaluative focus on "character traits, dispositions, emotions, and what I shall call 'virtues'" (Jamieson 2009, pp. 9–10). This is because, as I also argued, these can accommodate non-contingency, i.e., that an agent's actions are not "contingent on [her] beliefs about the behavior of others" (Jamieson 2009, p. 9) and are uncoupled from the actions of others (Jamieson 2009, p. 28).[9] However, Jamieson does not conclude from this that environmental ethicists should be virtue-oriented ethicists. Instead, because he endorses emulator utilitarianism, he concludes that "in this case and in this world, utilitarians should be virtue theorists" (Jamieson 2009, p. 28). In this way, Jamieson concurs with the conclusions argued for above, i.e., that the problem of inconsequentialism is a significant problem in environmental ethics and that virtue-oriented ethics is best resourced to address it, but disagrees that this implies virtue-oriented ethical theory is superior to utilitarian ethical theory, even with respect to such problems.

Because emulator utilitarianism and virtue-oriented ethical theory generate nearly identical normative theories, both structurally (i.e., regarding the structure of the normative theory) and substantively (i.e., regarding the content of the norms), which is preferable cannot be adjudicated on that basis,[10] but only on broader

[9] Jamieson also believes that a focus on character better positions virtue-oriented ethical theory with respect to amplifying effects responses: "Focusing on the virtues helps to regulate and coordinate behavior, express and contribute to the constitution of community through space and time, and helps to create empathy, sympathy, and solidarity among moral agents" (Jamieson 2009, p. 15). If these empirical claims are true, this is an additional strength of virtue-oriented ethical theory over act-oriented and duty-oriented theories for responding to longitudinal collective action environmental problems.

[10] This coincidence is possible because emulator utilitarianism is not a normative theory, it is meta-normative. It can, therefore, generate a virtue-oriented normative theory under particular conditions, while denying that such is the correct normative theory generally.

theoretical considerations. A fundamental point on which they come apart is their flexibility with respect to evaluative focal points. In virtue-oriented ethical theory, character traits are always the primary level of evaluation and actions are always evaluated in terms of virtues and vices. In emulator utilitarianism, the levels of evaluative foci and their relationships to one another can shift depending upon context. The proper level is the one that brings about the best consequences. Jamieson believes that this is an advantage of emulator utilitarianism over "indirect views that focus on motives, rules, or whatever" (Jamieson 2009, p. 12), which includes virtue-oriented ethical theory:

> All of these accounts are 'local', in that they privilege some particular 'level' at which we should evaluate the consequences of actions that are open to us. Rather than adopting any such local view, we should be 'global' utilitarians and focus on whatever level of evaluation in a particular situation that is conducive to bringing about the best state of affairs (Jamieson 2009, p. 12).[11]

However, there is a reason to want to privilege some level of evaluation over others. Doing so prevents pervasive irresolvable contradictory evaluations, which arise if no level of evaluation is privileged over any other. As Jamieson recognizes, due to its non-contingency, a virtue-oriented normative theory will often reach a different evaluative conclusion regarding an action than does act utilitarianism. This is why he advocates being a virtue-oriented ethicist and not an act utilitarian with respect to longitudinal collective action environmental problems. But on emulator utilitarianism, it is utilitarianism that is operating to justify virtue-oriented

[11] Jamieson's other primary argument against indirect theories (such as virtue-oriented ethics) is that they cannot support non-complacency: "Non-complacency refers to the fact that ways of life and patterns of action should be dynamically responsive to changing circumstances, taking advantage of unique opportunities to produce goodness, and always striving to do better" (Jamieson 2009, p. 15). It is important, for example, that one-off situations where acting contrary to a norm (e.g., virtue) would bring about great goods are accommodated by the theory, i.e., that it would not be wrong (and would be right) for an agent to take advantage of such situations. It is also important that the theory have a positive evaluation of "striving to do better," even if an agent is acting good enough. Virtue-oriented ethical theory can accommodate non-complacency. As discussed above, it is an open question, *qua* virtue-oriented ethical theory, how demanding a particular virtue-oriented ethical theory is, and ideals of character and action can be set that exceed that of those who to a substantial extent are already virtuous and acting well (Sandler 2007). Moreover, one-off situations where there is a great good to be gained only at the expense of acting contrary to some virtue will be cases of conflicting virtues. There will be some virtue operative in the situation that is more salient in that situation and that favors performing the action. Moreover, if such cases are common enough, sensitivity to them will inform the substantive content of the virtues (which, after all, are evaluated consequentially or teleologically). To the extent that such one-off cases arise within the domain of particular virtues, those virtues will involve appropriate responsiveness to them.

It is also worth noting that Jamieson's argument is not likely to resonate with many neo-Aristotelian virtue-oriented ethicists, even ones that are teleological in their evaluation of character traits (so committed in that sense to realizing or bringing things about), since they need not be committed to the point of ethics being realizing "the best state of affairs." Neo-Aristotelian ethics might, instead, ground ethics in human goodness or flourishing (Sandler 2007; Foot 2001; Hursthouse 1999). In this way, its meta-normative commitments would privilege (teleological) character level evaluation, even in those contexts where that level of evaluation does not bring about the best consequences (in the utilitarian sense). Jamieson might respond that this only betrays the rigidity of neo-Aristotelian ethical theory, and thus bolster the case that utilitarians should be "Aristotelians" sometimes but not others. Yet, again, this argument will only gain traction if the point of ethics is to bring about the best (or a good enough) state of affairs, which is an issue beyond the scope of this article.

normative theory. There is thus a conflict, with respect to what action is right, between emulator utilitarianism operating on the act level and emulator utilitarianism operating on the theory level. Which level takes precedent? Within emulator utilitarianism there is no theoretical basis for privileging either level. Nor is there any practical basis for privileging either level. Emulator utilitarianism is committed to bringing about the best consequences. But, given act level evaluation it would bring about the best consequences to perform one action (the one with greater local utility), while given normative theory level evaluation it would bring about the best consequences to follow a norm that favors a contrary action (the one that contributes to addressing the problem).

Jamieson might respond that the reason for privileging a theory level evaluative focus is that this level is effective in addressing longitudinal collective action environmental problems, whereas an act level evaluative focus is not. However, this only shows that one should not adopt act utilitarianism as a normative theory with respect to such problems. It does not show that, when deciding how to act in a particular instance, it would not be better to act in accordance with act utilitarianism. In fact, it would be better in terms of bringing about the best state of affairs in that instance. The reason for this is that any time one acts in a way relevant to a longitudinal collective action environmental problem, one is also just plain acting. This is the source of the tension within emulator utilitarianism. Simultaneous with supporting virtue-oriented evaluation of actions in order to address the problem of inconsequentialism, it supports equally well act utilitarian evaluation of actions that gives rise to the problem. When one asks, "from the perspective of emulator utilitarianism, should I evaluate this act as a virtue-oriented ethicists or an act utilitarian"'the answer is "both" or "either." They are equally theoretically and practically justified; they both bring about the best consequences (one from the normative ethics perspective, the other from the act perspective). There is thus a conflict between two levels of utilitarian evaluation, neither of which is more basic. This is not merely a methodological or implementation problem, it is a significant theoretical problem. Because of it, environmental ethicists should not be emulator utilitarianisms that emulate virtue-oriented ethical theory. They should be virtue-oriented ethicists.

Global utilitarianism differs from emulator utilitarianism in that global utilitarianism has several levels of evaluation—actions, policies, character traits, etc.—that are distinct from each other. An action is right to the extent that it brings about the best consequences. A character trait is a virtue to the extent that it brings about the best consequences. A policy is right to the extent that it brings about the best consequences. And so on. Unlike with emulator utilitarianism, in global utilitarianism the structure and foci of normative evaluations never change. All levels are always evaluated, and they are always evaluated directly in terms of consequences. As Derek Parfit has put it:

> Consequentialism covers, not just acts and outcomes, but also desires, dispositions, beliefs, emotions, the color of our eyes, the climate and everything else. More exactly, C covers anything that could make outcomes better or worse. (Parfit 1984, p. 25).

Therefore, on global utilitarianism it is never the case that actions are evaluated in terms of the virtues, even when the virtues are those character traits the possession of which bring about the best consequences. As Julia Driver has put it, "Character traits are simply another thing that, like action, can be evaluated along consequentialist lines" (Driver 2001, p. 72).

Global utilitarianism does not give rise to the contradictory evaluation problem generated by emulator utilitarianism. The problem arises for emulator utilitarianism because, with respect to longitudinal collective action environmental problems, evaluating acts according to act utilitarianism and evaluating acts according to virtue-oriented ethics are equally justified. In contrast, global utilitarianism always only evaluates actions according to act utilitarianism. However, due to the independence between character evaluation and act evaluation within global utilitarianism, it is subject to the problem of inconsequentialism in the same way as act utilitarianism, despite the fact that it involves consequentialist evaluation of character traits. (That is, it fails to exhibit features (1), (4), and (5) of a theory well resourced for responding to the problem on inconsequentialism, even as it exhibits (2) and (3).) It often is the case that the local utilities associated with an action that is complicit with a longitudinal collective action problem outweigh the (inconsequential) non-local utilities.

Global utilitarianism is perhaps a bit better off than act utilitarianism. Since it evaluates character traits in addition to actions, it can support counter-factual claims of this sort: if the agent were non-materialistic (i.e., virtuous) then it would not be right for her to consume so much. Nevertheless, because character and act evaluations are independent, on global utilitarianism it will still be right for the materialist to act consumptively if, due to her materialism, there are high local disutilities associated with acting non-consumptively. The rightness of her action is informed by her materialism. In contrast, virtue-oriented normative theory evaluates actions according to what a person's dispositions ought to be (i.e., in terms of what is the virtuous thing to do). Thus, global utilitarianism fails to address several core aspects of the problem of inconsequentialism.

Conclusion

Longitudinal collective action environmental problems are not the only environmental problems, but they are among the most significant. Virtue-oriented ethical theories are better resources for justifying individual responsiveness to such problems than are standard version of Kantian and non-virtue-oriented utilitarian theories. If providing a compelling response to the problem of inconsequentialism is an instance of theoretical or practical adequacy for environmental ethics, then this is a reason for preferring virtue-oriented ethical theory over those other ethical theories. This is not an all-things-considered argument for virtue-oriented ethical theories. There may be reasons, independent of the considerations discussed here, for preferring other approaches to normative ethics. But if the arguments above are correct, then environmental ethicists should be virtue-oriented ethicists, rather than non-virtue-oriented utilitarians or Kantian ethicists, all other things being equal.

Acknowledgments I thank Philip Cafaro, John Basl, Ben Miller, two anonymous referees, and participants at the 2008 meeting of the Inland Northwest Philosophy Conference for their helpful comments on earlier versions of this article.

References

Driver, J. (2001). *Uneasy virtue*. Cambridge, UK: Cambridge University Press.

Foot, P. (2001). *Natural goodness*. Oxford, UK: Oxford University Press.

Herman, B. (1993). *The practice of moral judgment*. Cambridge, MA: Harvard University Press.

Hooker, B. (2000). *Ideal code, real world*. Oxford, UK: Oxford University Press.

Hursthouse, R. (1999). *On virtue ethics*. Oxford, UK: Oxford University Press.

Jamieson, D. (2009). When utilitarians should be virtue theorists, *Utilitas*, preprint. Online at http://www.utilitas.org.uk/pdfs/jamieson.pdf. Accessed 18 Feb 2009.

Johnson, B. (2003). Ethical obligations in a tragedy of the commons. *Environmental Values, 12*, 271–287.

Kant, I. (1997). *Groundwork of the metaphysics of morals* (M. Gregor, Trans.). Cambridge, UK: Cambridge University Press.

Kasser, T. (2002). *The high price of materialism*. Cambridge, MA: MIT Press.

Kasser, T., & Kanner, A. (Eds.). (2003). *Psychology and consumer culture: The struggle for a good life in a materialistic world*. Washington, DC: American Psychological Association.

Korsgaard, C. (2004). *Fellow creatures: Kantian ethics and our duties to animals*. University of Michigan, Tanner Lectures on Human Values. Online at http://www.tannerlectures.utah.edu/lectures/documents/volume25/korsgaard_2005.pdf. Accessed 18 Feb 2009.

McDowell, J. (1980). The role of *Eudaimonia* in Aristotle's ethics. In A. O. Rorty (Ed.), *Essays on Aristotle's ethics* (pp. 359–376). Berkeley, CA: University of California Press.

Mill, J. S. (1972). *Utilitarianism, on liberty, considerations of representative government*. London, UK: J. M. Dent and Sons.

Parfit, D. (1984). *Reasons and persons*. Oxford, UK: Oxford University Press.

Regan, T. (1983). *The case for animal rights*. Berkeley, CA: University of California Press.

Sandler, R. (2007). *Character and environment*. New York, NY: Columbia University Press.

Swanton, C. (2003). *Virtue ethics: A pluralistic view*. Oxford, UK: Oxford University Press.

Westra, L. (2005). Virtue ethics as foundation for a global ethic. In R. Sandler & P. Cafaro (Eds.), *Environmental virtue ethics* (pp. 79–91). Lanham, MD: Rowman and Littlefield Press.

J Agric Environ Ethics (2010) 23:185–206
DOI 10.1007/s10806-009-9189-y

ARTICLES

Patriotism as an Environmental Virtue

Philip Cafaro

Accepted: 27 May 2009 / Published online: 17 June 2009

Abstract Define "patriotism" as love for one's country and devotion to its well-being. This essay contends that patriotism thus defined is a virtue and that environmentalism is one of its most important manifestations. Patriotism, as devotion to particular places and people, can occur at various levels, from the local to the national. Knowing and caring about particular places and people and working to protect them is good for us and good for them and hence a good thing overall. Knowing and caring and working less on behalf of more remote places and people is also good, since it allows us to focus our efforts, act effectively, and do more good in the world. Philosophical analyses of patriotism by Alasdair MacIntyre and Martha Nussbaum are complemented by the more "down to earth" understanding of the virtue presented here. While patriotism's dangers are undeniable, so are the dangers stemming from lack of patriotism. The proper answer to bad patriotism is not cosmopolitanism, but good patriotism: the kind illustrated by environmental activists.

Keywords Virtue · Patriotism · Cosmopolitanism · MacIntyre · Nussbaum · Thoreau

Introduction

I think that patriotism is a good thing. Or rather, I think that *the right kind of patriotism* is *usually* a good thing. Patriotism, like all good things, can go wrong in various ways and wind up doing more harm than good. I don't mean to ignore patriotism's dangers, and address some of them below. Meanwhile, putting my main thesis into what I hope is a slightly clearer form, I assert: "Patriotism is a virtue."

P. Cafaro (✉)
Philosophy Department, Colorado State University, Fort Collins, CO 80523, USA
e-mail: philip.cafaro@colostate.edu

What do I mean by patriotism? My dictionaries' primary definitions of "patriotism" are "love of and devotion to one's country" and "devotion to the well-being or interests of one's country" (Soukhanov et al. 1992; Brown et al. 1993). These definitions accurately capture current usage. However, older understandings of "patriot," "patriotism" and even "country" were more flexible, allowing for patriotic connection and devotion to various particular localities and groups, at smaller scales than the nation-state (see entries in the unabridged *Oxford English Dictionary*). I find this older understanding of the term appealing, in recognizing the range of patriotic connections open to individuals, and the possibility of less abstract, more "grounded" forms of local patriotism.

Let me stipulate then that in what follows, by "patriotism" I will mean love, devotion, and a strong differential concern for one's own locality, state, region, or country, shown both in thought and action.[1] When I say "concern," I mean concern for both the places themselves and the people who live within them. When I say "a *differential* concern," I mean you care more about these places than you do about other places, or places in general; and that you care more about these particular people than you do about other people, or people in general. And this concern is shown in substantial activity on their behalf.

A patriot is particularly concerned to defend his own country and countrymen and women, however, broadly or narrowly he defines these, and to promote their well-being and interests. He might condemn an imperialistic war between two foreign countries and boycott the aggressor country's goods. But he will risk his life to defend his own country from attack. A patriot might condemn the exploitation of poor laborers in third-world sweatshops, and sign a petition asking Nike to change. But she will care more about poor people in her own community and spend some of her own valuable time to improve their lives (work one evening a month in a soup kitchen; testify before the city council in support of affordable housing). This differential attentiveness, concern, and action are what I take to be patriotism.[2]

What do I mean by virtue? "Virtue" is the generic term commonly used for any character trait people wish to commend. In both common speech and philosophical discourse, "the virtues" refer to those traits whose possession we believe makes a person a good person, or a better person than she would be without them. They do this by helping people succeed in what we take to be our characteristic or important activities and endeavors.

For the most part, current philosophers endorse one of two different conceptions of virtue. On the one hand, there are moralistic conceptions, such as Julia Driver's: "a moral virtue is a character trait which, generally speaking, produces good

[1] Miller (1998) usefully distinguishes differential *concern* from differential *valuation* of people: "on the face of it, according equal value to different people's lives does not entail equal concern for them" (p. 207). *Pace* Gomberg (1990) and other critics, patriotism involves greater concern for compatriots, but not necessarily a belief in their superiority.

[2] It is of course possible to help people locally or (less often, perhaps) work to protect local places, based on strongly held universalistic moral principles. So every instance of locally-focused altruism does not constitute patriotism, as I've defined it. On the other hand, it would be a mistake to define patriotism as action taken with no reference to such universal moral principles or ideals. People often appear to be motivated by particular loves *and* universal principles, in one and the same commendatory act.

consequences for others" (Driver 1996, p. 122). Here virtue is explicitly altruistic: good because it secures good consequences for others, not for the virtuous individual herself. Driver offers this as a definition of *moral* virtue, but she recognizes no other kind. Virtue simply is moral virtue. Excellent people are people who act morally.

On the other hand, there are conceptions of virtue that equate it with human excellence more comprehensively understood. For example, Martha Nussbaum defines the virtues as character traits which dispose us to "choose and respond well" across the full range of human experience (Nussbaum 1993, p. 245). In this way, the virtues promote the flourishing of the virtuous agent herself, as well as the flourishing of those around her. This expanded notion of virtue supports a broader list of virtues. On this view, the virtues may include *moral* virtues such as kindness and justice and generosity, *personal* virtues such as self-knowledge, integrity and commitment, and *intellectual* virtues such as patience, intelligence and clarity of expression. Because there is more to life than living morally, I think this more expansive view of virtue deserves to prevail, and follow it below.

However, even this broad view of virtue needs broadening, because just as human flourishing involves more than moral excellence, it also involves more than *human* excellence. We depend on nature. In particular, our continued existence depends on maintaining key ecosystem services (Millennium Ecosystem Assessment 2005) and our full flourishing depends on preserving sufficient wild, untamed nature for our physical, intellectual, and spiritual sustenance (Thoreau 1971; Leopold 1949). Arguably, virtue also involves recognizing the intrinsic value of nonhuman beings and responding accordingly (Rolston 1988; Swanton 2003, pp. 46, 94). For all these reasons, my own list of virtues includes *environmental* virtues such as temperance, stewardship, and respect for nature: qualities that further ecological sustainability, the necessary grounding for human flourishing (Wensveen 2001). Such environmental virtues have recently been usefully explored and typologized by Sandler (2007). Sandler, however, forgot to include patriotism among them (hence this article).

The virtues, then, on my view, are *all* those qualities that help us perform characteristic, essential, or important human activities well. They thus facilitate personal, social, and ecological flourishing—the flourishing of all life. I believe this, ultimately, is what makes them virtues (Cafaro 2004, pp. 45–65). *Vita qua vita bonum est.* It is sometimes necessary to distinguish individual, social, and ecological flourishing for analytic purposes, particularly when they conflict, or appear to conflict. But in my opinion, all three are so supremely valuable and so intimately related that no human character trait that undermines any one of them counts as a genuine virtue.

Once again, this expansive definition of "virtue" differs from common usage and much philosophical usage, which often limits virtue to moral virtue, narrowly understood. Since I intend to argue that patriotism *is* a moral virtue—a character trait that promotes altruism and helps us act rightly toward other people—as well as an environmental virtue, you may disagree with my broad definition of virtue and still agree with me that patriotism is one. But the broad definition is important to fully understanding my argument, which is that patriotism is good for nature, good for human societies, and good for us as individuals—this last, despite the fact that it sometimes does involve self-sacrifice.

Why Patriotism is a Virtue

Patriotism is a virtue in part because it sustains human communities, which are essential to human flourishing. The ancient Greeks had no trouble understanding this. They recognized that the *polis* could be fatally endangered. Their land could be taken away from them, their families enslaved by aggressive outsiders. The main defense against this was their citizens' patriotic willingness to risk their lives to defend their homelands.

Even in a country like the United States, with all our wealth and power, our safety depends on the willingness of certain individuals to risk their lives for the rest of us. If most of us are unlikely to be called to sacrifice our lives for the greater good, some are much more likely: soldiers, firefighters, policemen working in dangerous neighborhoods. It's not clear that this division of labor (a huge protected majority, a small minority of brave protectors) is good for our country as a whole. But the simple point I want to make here is that even in individualistic America, we depend on some people being self-sacrificing and patriotic—whether we are or not.

This example helps define a basic, minimal need for patriotism. As things currently stand, without people willing to die for their country, a country cannot survive. What I want to say now is that patriotism also demands that citizens *live* for their countries. A less heroic patriotic concern and effort are needed, so that our nations and our communities may not only survive, but flourish. Such effort is shown in social and civic-minded projects of all kinds: volunteer fire departments; efforts to create new parks; volunteering in local schools or on local school boards. These take ongoing effort. The health of our communities depends, in part, on how many people take the time to engage in them. Without these efforts, houses burn, children do without parks, and schools fall into mediocrity. Without them, our communities are worse communities.

Engaging in such activities usually demands sacrifices, in time and money. But they also typically provide real benefits to the individuals making the sacrifices. Here we come to a second main reason why patriotism is a virtue: it is good for us as individuals, in several ways.

First, patriotic activities help us feel connected and affiliated with other people. They introduce us to our neighbors and fellow citizens, and these connections make life more interesting and enjoyable. Second, they lend meaning to our lives, as we engage in projects that are larger than ourselves. It's good to feel that you've had some say in how your town is governed, to point to a park or greenway you've helped create, or to a school that you've helped improve. Conversely, it feels bad to walk through the town where you live and feel like an outsider, unconnected to the people and places around you.

Now, this is not to say that becoming socially involved is always wonderful. Local politics is often tedious. Sometimes you get to know the neighbors and you wish you hadn't. But most of us have enough choice in how to get involved that on balance, the personal connections and enhanced meaning we get from these activities improve our lives. We miss out on a lot when we withdraw from political and social engagement, and our communities miss out, too.

Here I'd like to emphasize again that patriotism, as I define it, is not coextensive with nationalism. If we are lucky, we have numerous affiliations that give our lives meaning, and belong to various communities that provide scope for our efforts (Appiah 1996). For me, these include Fort Collins and Colorado State University and the short-grass steppe ecoregion, as well as Colorado and the United States. Patriotic connection and effective action are often easier to make sense of at more particular, local levels. Nothing has done more to distort our understanding of patriotism than the facile equation of patriotism and nationalism (Pinsky 1996).

The ancient Greeks had an easier time of it. Living in small city–states, their patriotic limits were simpler and more clearly defined. Even then, however, clashes of affiliation existed; witness the conflicting claims of family and *polis* in Sophocles's *Antigone*. We too may face conflicting claims on our allegiance. But I think that before we can say anything very intelligent about them, we need to get a much fuller view of our patriotic possibilities.

It is also true that the local efforts I'm praising can be undertaken from a variety of motivations, including general feelings of benevolence; and justified for a variety of reasons, such as their upholding universal moral principles, or helping to maximize overall happiness. Still, I think that attention to the phenomenology of such activities will show the importance of particular connections, for most participants. I care how well *this* class of second-graders is reading, or whether or not *this* natural area is paved over to build a new mall. That is why I take action. Such particular concern increases the likelihood of engagement. It helps us engage effectively. It enhances the meaning of these activities for the participants (perhaps by connecting general ideals or principles to lived experience and to our interactions with people and places we actually know). Particular connections also keep "doing good" from becoming just an impersonal transfer of resources and instead help make such activities "good in themselves." Patriotic particularism thus increases the good generated by our do-gooding.

Meeting the Dangers of Patriotism

I have said a little about the benefits of patriotism. What are its dangers?

The first and greatest danger is contempt or (more often) indifference toward outsiders. Our tender feelings toward our fellow citizens can make it easier to dismiss the value of people outside this charmed circle, or at least act in ways that fail to honor that value. Patriotism can encourage an "us versus them" mentality, and rationalizations for "our" superiority are easily generated.

A second danger is the stifling of dissent within our own communities. Criticism of the community may be seen as unpatriotic and uncritically following orders may be seen as the highest form of patriotism. When these two aspects of patriotism—contempt or indifference toward outsiders, the stifling of dissent within a community—synergize at the national level, then imperialistic aggression and war become more likely. I think it is fair to say that patriotic feelings and uncritical patriotism have sometimes facilitated unjust wars.

These dangers of patriotism are real. How should we address them? Here politicians and political thinkers on both the left and the right are often unhelpful. Liberals, by and large, have abandoned patriotism, perhaps believing its dangers are not worth its benefits (but see Rorty 1998, for a plausible attempt to specify a liberal patriotism). Conservatives, contrarily, too often embrace patriotism uncritically, saying, in effect: "wave your flag, support your government, and shut up."[3]

The insufficiencies of both approaches are obvious. In particular, both undermine political involvement. For without some sort of patriotic concern, I have little reason to get involved in politics (beyond private interest, which will rarely get the average citizen involved). Abstract ideals may motivate political action, but ideals are more effective when combined with a concern for the well-being of particular communities toward which people feel strong affiliation. And on the other hand, blind obedience is worse than no political involvement at all.

Rather than follow either of these paths, I suggest that we define and practice a better sort of patriotism. We can loyally work to protect the places and people within our particular circles of concern, without dismissing the value of those outside these circles, or aggressing against them. Perhaps we can see our way clearer to such a position by recognizing a certain arbitrariness, or chance, in who or what we take to be inside or outside the circle! I was born an American and some combination of chance and my own decisions have led me to live in northern Colorado. So I am more concerned with American foreign policy and the Roosevelt National Forest and homeless people in downtown Fort Collins, than I am with French foreign policy or the Coconino National Forest in Arizona or homeless people in Paris or Tucson. But that doesn't mean I think these other people and places are any less important or valuable, considered absolutely.[4] That doesn't mean I will throw trash on the Coconino when I go hiking there, or kick a homeless person when I'm walking around Tucson or Paris.

One key to curbing the dangers of patriotism, then, is to recognize that basic moral rules still hold, regarding how we may treat other people and places. Anything *doesn't* go, once we are outside our circles of patriotic concern. A basic, universal morality remains (Nathanson 1989).[5] I see no reason why this must fatally conflict with patriotism, any more than a special concern for our family

[3] Conservatives also often use patriotic symbols as means to score cheap political points; witness the flap, during the 2008 US presidential campaign, over Barack Obama's missing flag pin. Such episodes lend credence to charges that patriotism is only for dolts and those who want to control them.

[4] Here I disagree with McCabe (1997), who writes: "patriotism requires… that one believe one's nation is better, and for this reason deserves more" than other nations (pp. 207–208).

[5] I thus substantially agree with Nathanson (1989) when he writes: "the proper answer to the question, Is patriotism a virtue? is that the moral value of patriotism depends on the circumstances in which patriotism is exhibited and the actions that it motivates. When patriotism is in the service of valuable ends and is limited to morally legitimate means of attaining them, then it is a virtue. When patriotism leads to support of immoral ends or immoral means to achieve otherwise legitimate ends, then it is a vice" (p. 539). If I understand Nathanson correctly, however, I am unlike him in leaving it an open question, whether universal or particular moral commitments should prevail, in particular instances when they conflict. See Schmidtz (1997) for a searching discussion of this issue.

members must do so. This is particularly true in Western democracies, where an essential part of the political legacy is a commitment to universal human rights (Barber 1996).

Of course, occasional conflicts between universal morality and our special areas of loyalty will occur, and these may be hard to adjudicate. But most of the time these values enhance one another. This is because our particular loyalties provide the spheres of activity where we can best work to uphold morality. We show our patriotic concern by advocating for more morally acceptable policies from our government, our familial concern by cultivating moral behavior in our children. It is true that resources are limited and $200 given to a local political candidate is $200 not sent to Oxfam to provide family planning or education in less developed nations. But such conflicts are unavoidable; ignoring our particular loyalties only simplifies this situation by ensuring that we will choose wrongly (by failing to take care of our own children's needs before sending money to Oxfam, for example).

The second key to defining a good sort of patriotism is to recognize the value of patriotic dissent. Real democracies are participatory democracies, which depend on citizens intelligently debating important issues. Sometimes we will agree with our government or with a majority of our fellow citizens, sometimes we will not, and it is probably more important that we voice our concerns when we are in the minority. American history is filled with examples of patriotic dissent, from Thomas Jefferson's criticisms of John Adams's Alien and Sedition Acts, to Martin Luther King, Jr.'s civil disobedience in Birmingham. No important political progress has been made in this country without intelligent dissent to challenge the moral laziness and mammon-worshipping tendencies of Americans.

Let me make this point a little clearer. We easily—perhaps too easily—recognize the patriotism of the men and women who serve in our armed forces. We also need to see the patriotism of people who help us think harder about how those armed forces should be deployed. Ninety-nine members of the United States Senate voted to pass the USA Patriot Act a few weeks after 9/11; we need to recognize the courage and patriotism of the lone member who voted against it (Russell Feingold, D-Wisconsin). Most important, we need to see patriotism in a thousand less heroic actions that help sustain society.

Still, you might say that given its dangers, we'd be better off without patriotism. I've suggested, first, that a special concern for our own communities and nations helps sustain them and us. And second, that patriotism's dangers need to be balanced by an understanding of the dangers stemming from a lack of patriotism. For Americans, at the present time, I judge that the dangers of political apathy and social fragmentation outweigh the dangers of excess patriotic zeal (McConnell 1996; Putnam 2001). I leave it to patriots in other countries to make their own judgments here.

My third reply supporting "constructive engagement" with patriotism is that some kinds of particular affiliation are inevitable, so we might as well support the right kinds. Human beings naturally seek community and find meaning in projects larger than themselves. We can do this more or less effectively and intelligently. Mock all patriotic concern, and you will be met with "America: Love it or Leave it." You will find it harder to talk to your fellow citizens about the common good, while political reactionaries and corporate lobbyists step in to define it for you. For all these reasons, I believe a proper patriotism is both possible and necessary.

Environmentalism as Patriotism

Let me relate this to a special concern of many of us: environmental protection. In my view, environmentalism, rightly understood, *is* patriotism.[6] The word comes from the Latin root "patria": fatherland or homeland. So patriotism would seem to indicate a general concern to protect the land and the people of one's homeland, however extensively one defines them. On this view, environmentalists are patriots *par excellence*, literally committed to protecting the land and *all* its inhabitants, including its non-human inhabitants.

Environmental activists typically work to protect the places they know and love, whether it is open-space threatened by sprawl, or a downtown threatened by a new Super Wal-mart. They involve themselves in their communities: leading hikes to teach boy scouts and girl scouts about the plants and animals around them, or tipping off reporters to illegal dumping. They hand out campaign literature for green candidates and pack city council meetings to argue about zoning changes. They do all this not to make a buck, but because they care about where they live and other special places they have gotten to know, and because they want to preserve them for their children and grandchildren.[7]

Environmentalists don't only protect these places, they *celebrate* them as well: hiking and camping and kayaking and skinny-dipping to show their patriotic attachment to their country. Just as a 4th of July parade with marching bands provides a sense of belonging and larger community to many, so may a hike up Long's Peak, or an Audubon Society field trip to watch the birds at a local reservoir. In these activities, people bond with the land and with each other. They come face to face with what will otherwise remain mere abstractions: biodiversity; "America the beautiful." They learn the history of the natural and human communities of which they are parts. They come to see their own lives as parts of larger stories, their own efforts as parts of larger struggles and achievements.

This is patriotism! And it is the kind of patriotism that we need more of today. We live in a country whose citizens move on average every 6 years; and where more and more people spend the best part of their days gazing into computer screens, focused on virtual realities of one kind or another. In this world, those who work to strengthen the bonds of community and our ties to the real, physical world around us are doing necessary, patriotic work (Nelson 2002).

---

[6] My understanding of the connection between environmentalism and patriotism has been greatly influenced by Nelson (2002).

[7] As applied to the United States, this locally-focused description of "environmentalists" leaves out some top leaders and staff members of large environmental organizations working primarily at the national level, and a few thousand dedicated souls focused on issues far from home (sending high-tech cooking stoves to poor people overseas, for example, to cut back on deforestation and indoor air pollution). But the description does capture the vast majority of serious environmental activists: the backbone of the environmental movement.

We also live in a world where large corporations can crush a community or destroy a landscape with frightening ease. Since many corporate executives will attempt to wring the last dollar out of a landscape or a community with no moral scruples, our land and our communities need patriots who will identify and defend the common good against these behemoths. (When they do so, they are often derided as "nimby's"; short for "not in my back yard." As if indifference to our communities and surroundings were the height of rationality. The proper name for "nimbyism" is patriotism.)

In fact, appeals to patriotism have been important and effective in many environmental battles over the years. In the campaigns to create Yellowstone, Yosemite, Grand Canyon, and other national parks in the nineteenth and early twentieth centuries, patriotic rhetoric often figured prominently (Runte 1979). Park boosters appealed to citizens' desire to preserve what was grand or unique about America, or to preserve historically meaningful landscapes.

Again, when the United States enacted foundational environmental laws like the Clean Air Act and Clean Water Act in the 1960s and 1970s, these were often promoted in overtly patriotic terms. Preserving a healthy landscape was described as farseeing and patriotic; polluters were not just harming innocent people, they were putting profits over country. Perhaps it isn't an accident that the last time environmentalists won major victories, in the early 1970s, was also the last time such patriotic rhetoric was widely and unironically employed by the American left.

Particularism Versus Universalism

The kind of patriotic commitment I'm advocating, however, demands focus. Moral philosophers often develop universal arguments about the equal value of all X's or Y's. "All people have certain inalienable rights." "All sentient beings have intrinsic value." These arguments are necessary, I think, to help articulate a correct ethics. But they mislead us if we assume that they show that we should, or can, *care equally* about all X's or Y's, or that we *have equal responsibility* for all X's and Y's, or that we should *devote equal time and resources* to trying to protect all X's or Y's (Walzer 1983; Miller 1995).

If I am going to be an effective environmental advocate, I have got to focus. I can say something intelligent about conservation issues in northern Colorado, because I've learned a little about the land and its history, and the potential effects of particular land-use changes. I can advocate more effectively here because I know some of the land managers and the politicians in the area. I can call up editors and reporters on the local newspapers and pitch stories, or try to influence their coverage of important ongoing issues. The mayor of Fort Collins may be a pro-growth Republican, but I know that he grew up trout fishing the local Cache la Poudre River with his father. We've talked about what that meant to him and what my own rambles along the river with my sons mean to me. Maybe that made a difference recently, when he voted with the greenies on city council to oppose a new dam project that would harm the river.

Similarly, I know and care more about Yellowstone National Park than about Krueger National Park in South Africa, and this too allows me to more effectively advocate for its protection. As a former ranger with the National Park Service, I can call up park managers and discuss their upcoming decisions more easily. As an American citizen, I can pester my Congresswoman's staff about bills that affect Yellowstone, or write opinion pieces in the newspaper that alert my fellow citizens to problems there. Such actions are more likely to do good than signing on-line petitions to the chief of Krueger National Park.

Now, this doesn't mean we should be completely indifferent to what is happening further afield. In fact, a person with a deep knowledge of environmental issues in northern Colorado can better understand what is happening in southern Colorado or Utah; a person familiar with the issues facing US national parks can better understand management issues in national parks around the world. He or she is also more likely to explore other places respectfully and support international conservation efforts. But all of this usually begins at home!

So does our *appreciation* of nature (Louv 2005). I can only enjoy and celebrate those aspects of my country that I experience personally. I might get a thrill hearing Ray Charles sing "America the Beautiful," but it really means something to me because I have recently seen a sunrise over the Pawnee Buttes east of Fort Collins, or the first bluebird of spring on a friend's ranch. I can enjoy nature and belong locally, in a way that I cannot do elsewhere. And environmentalism can only be a life-affirming and personally enriching activity if it involves connection to the land and communities around you.[8]

A positive local patriotism is the key to environmentalists avoiding the Scylla of indifference and the Charybdis of burnout. Enjoy nature, in whatever way most engages you. Learn the plants, the birds, the insects, the stars—and teach them to your children, your students, your friends. Learn the history of where you live (its real history, warts and all) and resolve to write a new chapter to the local story of which you can be proud.

The Meaning of America

Of course, we are not just residents of Fort Collins or Baltimore or Butte, but also, many of us, Americans. At this juncture in history, I think that gives us some unique opportunities, both at home and abroad.

The United States has contributed two great political innovations to the modern world. The first is a Constitution that guarantees individual rights to all its citizens. The second is our system of national parks: lands set aside to protect a wild heritage

[8] "Who cares," you might ask, "whether environmentalism is life-affirming and personally enriching? Environmentalists' goal should be to convince people to behave in more environmentally responsible ways—not to make people feel good." I answer, first, that if doing good doesn't feel good, or isn't part of larger changes in our lives that improve them, we are less likely to do good. Many environmental examples could be given. Second, that on my view, virtue proves itself by furthering the flourishing of life, including the flourishing of the virtuous person's own life. For a general defense of such an environmental virtue ethics, see Cafaro (2005).

for future generations. Both these innovations have been widely copied abroad; both are works in progress here at home. They point to ideals of human dignity and human restraint that the world's people should strive to further actualize.

I believe that our current environmental conflicts are more than struggles between different interest groups. Rather they are struggles about the meaning of America. What will future historians identify among our most lasting contributions to world history? National parks and the world's first important international environmental treaty (the Pan-American Migratory Bird Treaty of 1904)? Or our country's undermining world efforts to combat global warming, under the Kyoto accords? Will history celebrate America's trailblazing commitment to preserving all the native species under our stewardship, with the Endangered Species Act? Or will it remember instead our gluttonous consumerism, which undermines human and ecosystem health worldwide? Our actions will answer these questions. American environmentalists need to convince our fellow citizens that we have the nobler, more generous conception of what our country should be (Nelson 2002). We need to appeal to their patriotic concern to motivate them to action. We need to remind them that in addition to the dominant materialistic tradition, there is an American tradition of "plain living and high thinking" that needs revival (Shi 1985).

Patriotism can be powerful. This is yet another reason why it is a mistake to cede it to militarists and reactionaries. Eight years after 9/11, security remains a strong concern for many Americans (too strong, in my opinion). Those leaders who can speak convincingly to common security concerns will likely continue to find the way open for their favored policies. It is important for environmentalists to speak to our fellow citizens' security worries and articulate a compelling vision of the "national interest," so that in meeting them, we really benefit our country and the rest of the world.

For example: was it truly in our national interest to oppose international efforts to combat global warming, during the Clinton and Bush administrations? Will it really enhance national security to drill for oil in the Arctic National Wildlife Refuge and other pristine wilderness areas? The figures on US oil reserves and rising global temperatures suggest that promoting conservation and developing alternative energy sources are better bets for increasing our security and the security of the rest of the world (IPCC 2007).

Again: would a true patriot encourage snowmobile and jet ski use in our national parks? Rescind protection for streams and rivers threatened by mine-related pollution? Weaken the Endangered Species Act? I don't think so. If there was any patriotic concern for the land—the "homeland"—during George W. Bush's administration, it certainly took a back seat to corporate profits and paybacks to campaign contributors. Environmentalists, who generally skew left politically, might have been able to block some of these assaults on nature, if we could have found common cause with conservatives for whom patriotic themes resonate (Bliese 1997). If environmentalists had successfully branded Bush's anti-environmentalism as unpatriotic (which it was), we might have made it harder for him to stampede Americans into the Iraq war.

I believe Americans, today, are failing in our important moral responsibility to help sustain the global life-support systems on which all humanity depends. As an American and, I hope, a patriot, I take this failure to heart. Unlike President Bush, I will not say that it is your patriotic duty to shop for your country. Unlike President Obama, I do not agree that a growing economy is the *sine qua non* of American well-being. In fact, I believe it is our patriotic duty to consume less and to question more, in order to help move our country toward a just and sustainable future (McKibben 2007). A nation of pampered consumers will not be able to stand up for our highest ideals and act morally on the world stage. We can and must do better.

MacIntyre on Patriotism

Reconnecting patriotism to *patria*, the land, can help correct some errors in recent philosophical explorations of patriotism. The errors in the two cases that I explore below stem from a kind of over intellectualism. Specifically, their authors substitute an intellectual construct for the real warts-and-all phenomenon of patriotism, praising or demonizing it accordingly. So while their discussions may illuminate possibilities in ethical theory, their generalizations about patriotism itself are suspect.

In his influential lecture "Is Patriotism a Virtue?" (1984) Alasdair MacIntyre helpfully classes patriotism as a "loyalty-exhibiting virtue," along with marital fidelity, love of family, friendship, and personal commitment to various institutions. "All these attitudes exhibit a peculiar action-generating regard for particular persons, institutions or groups, a regard founded upon a particular historical relationship of association" (p. 308). "The particularity of the relationship is essential and ineliminable," MacIntyre affirms (p. 309). Marital fidelity as a virtue is not a commitment to marriage generally, but to my particular spouse; the virtue of patriotism does not play out as a commitment to citizenship or nationalism as ideals, but to my particular city or country. This seems accurate both to the meaning and the experience of marriage and patriotism.

MacIntyre goes on to discuss how patriotism, thus defined, necessarily conflicts with any plausible universal morality, including the impartial demands of justice. It is a defining feature of universal moralities that they may call into question all aspects of our lives, personal or communal, no matter how ancient, well-entrenched, beloved, or supposedly sacred. Such a universal acid, MacIntyre believes, necessarily conflicts with the robust acceptance of particular communal beliefs and commitments that is at the heart of patriotism. But the conflict works both ways. "While the liberal moralist (is) able to conclude that patriotism is a permanent source of moral danger because of the way it places our ties to our nation beyond rational criticism, the moralist who defends patriotism is able to conclude that liberal morality is a permanent source of moral danger because of the way it renders our social and moral ties too open to dissolution by rational criticism" (p. 317).

MacIntyre thus uses the issue of patriotism to articulate the theoretical concerns at the heart of *After Virtue*, including the fundamental incommensurability of different moral vocabularies, the "systematic incompatibility" of different moral standpoints (p. 309) and the need to ground morality in particular traditions and practices. Because of MacIntyre's *theoretical* commitment to this last position, "patriotism" for him will be a virtue. We cannot have any morality at all without some degree of "patriotism," thus defined. But it will be a patriotism where at least some of our particular commitments are off-limits to the full force of rational criticism. If it was not, MacIntyre believes, then it would not be patriotism, but simply another form of moral universalism.

Just here, I think, is where MacIntyre goes astray. For we may hold robust patriotic commitments, not just while also holding robust universal moral commitments (which he acknowledges). We may hold robust patriotic commitments while accepting that *all* our practices, *all* our particular relationships, the meaning of *all* our values, *all* our tentative weightings and balancing of those values, and *all* aspects of our communal narratives, are up for continuous rethinking and renegotiation (Dewey 1948). Indeed, today, taking patriotism seriously and acting upon it in a complex and changing world demands such rethinking and renegotiation.

Consider the American environmental patriot. She asks us to question and change practices that are unsustainable and in fact threats to our communities; even well-entrenched, beloved practices like personal automobility. She asks us, following Aldo Leopold, to rethink and strengthen our relationships with the nonhuman world. She questions fundamental American values such as individualism and endless economic growth, and asks us to recalibrate how we balance them against other values, such as sustainability and the strength of our communities. She asks us to reconsider national narratives that Americans find deeply meaningful, such as the "settlement" of our country, and write new chapters that take into account new ecological realities. All this calls for more thinking, not less. Such thinking is itself an important part of patriotism. And crucially, such thinking may call for *limiting* a community's or nation's acquisitiveness or aggression.

MacIntyre is skeptical of the possibility of a moderate or "liberal patriotism," which limits its patriotic ardor within the dictates of universalistic morality. Such a patriotism would be "emasculated," he suggests, because "important situations in actual social life" arise where "the patriotic standpoint comes into serious conflict with the standpoint of a genuinely impersonal morality or it amounts to no more than a set of practically empty slogans" (p. 309). And what are these important, actual situations? Interestingly, the only examples MacIntyre gives are international competitions for "essential resources," such as arable land or fossil fuels:

> What your community requires as the material prerequisites for your survival as a distinctive community and your growth into a distinctive nation may be exclusive use of the same or some of the same natural resources as my community requires for its survival and growth into a distinction nation. When such a conflict arises, the standpoint of impersonal morality requires an allocation of goods such that each individual person counts for one and no more than one, while the patriotic standpoint requires that I strive to further the interests of my community and you strive to further those of yours, and certainly where the survival of one community is at stake, and sometimes

> perhaps even when only large interests of one community are at stake, patriotism entails a willingness to go to war on one's community's behalf. (pp. 309–310)

Now *if* we frame the resource conflict as one of *survival* and *if* we posit two radically different moral views, patriotic particularism and ethical universalism, then MacIntyre is clearly right. These orientations will push us in opposite directions and patriotism will sometimes push us toward war. The example is analogous to similar ones designed to bring out the importance of other particular loyalties: the mother choosing to save her drowning son rather than a stranger, etc. The theoretical point is made.

But at a time when Americans have fought two oil wars within a dozen years, I think it is important to think about the actual causes of such resource wars. The first and second Iraq wars did not involve America's "survival," but arguably, some carefully obscured combination of corporate profits, strategic *realpolitik*, and perpetuation of a wasteful and ecologically destructive "way of life." Iraq's oil is not required for our "growth into a distinctive nation," although it may be required for the endless economic growth that most Americans accept as our great national goal. The environmentalist argues that this "way of life" and the goal of endless economic growth would be well lost, *from a patriotic perspective as well as from a universalistic one*. Our gluttony is harming the United States as badly as it is damaging the global ecosystem services on which all the world's peoples depend, and a searching consideration of our "national interest" will show this. Further, he argues that meeting the global environmental challenges of the twenty-first century will require unprecedented amounts of international cooperation. Again, from either a universalistic or a particularist perspective, war is now obsolete as a means to solve conflicts over scarce resources. The imperative of endless economic growth, to the extent that it makes resource wars more likely, must be decisively rejected, if humanity hopes to survive and prosper in this century and the next (Millennium Ecosystem Assessment 2005).

MacIntyre's analysis is not completely wrong. We can be sure that just as there will remain conflicting interests between nations, there will remain cases where particularist and universalist moralities push us in different directions. What a serious consideration of the environmental example shows is that we can think our way through to better, more ethically-justified positions in such conflict cases by thinking harder about our patriotic interests and values, and weighing them against universalistic interests and values that also remain available to us. It is not just immoral, it is foolish to try to take refuge in an unthinking patriotism here. It will not help us further our self-interest—any more than invading Iraq and squatting on its oil reserves have furthered America's self-interest. There is no magic equation that provides optimal solutions to such complex, conflicted situations. The best solutions, I submit, will be found by those who are aware of the full range of interests and values involved and subject them to searching scrutiny (Dewey 1948).

It is a deep mistake of many philosophers to equate "patriotism" with intellectual limitations.[9] This mistake occurs not just because intellectuals tend to look down on those who find patriotism meaningful and not just because we tend to live in our own countries of ideas, which may feel more real to us than our actual countries (clearly, MacIntyre avoids these failures). It also occurs because many philosophers still hanker after absolute certainty in ethics (Rorty 1998). Taking his analysis, correctly, up to the point of recognizing that patriotic loyalties and universalistic morality will sometimes clash, and realizing that *reason may not always be able to clearly adjudicate between them*, MacIntyre draws the moral that patriotism, itself, needs to take refuge in irrationality. *Erratum*. Patriotic particularism and ethical universalism both run up against the limits to rational justification in ethics, and so will the best combination of the two. Now and forever. That is what we have to learn to live with, whatever our theoretical preferences in ethics.

Nussbaum on Patriotism

Of course, most philosophers are considerably less sympathetic to patriotism than MacIntyre. In "Patriotism and Cosmopolitanism," Nussbaum (1996) accuses patriotic nationalists of substituting "a colorful idol for the substantive universal values of justice and right" (p. 5). "Nationality, ethnicity, religion, class, race, and gender" are all, she claims, "morally irrelevant" (p. 133). All human beings, simply as human beings, have equal moral value. Morality involves recognizing this universal value and acting upon it—often in the face of surface differences or particular loyalties that obscure it. The Nazis were ethical particularists and patriots, willing to die to further the interests of the Fatherland, or kill to keep the Aryan race pure. In contrast, the "righteous Gentiles" memorialized at Yad Vashem, the Holocaust memorial in Jerusalem, were ethical universalists, willing to risk death to save their fellow human beings (pp. 131–132).

If all patriots were Nazis and all universalists were truly righteous, this would be the end of the story. But as Walzer (1996) points out, in a commentary on Nussbaum's essay, cosmopolitanism has its pathologies, too. The Nazis must share pride of place as the great butchers of the twentieth century with Russian and Chinese communists, followers of a universalist ideology. Patriots, meanwhile, may do much good. The Nazis were ultimately defeated by patriots from England, Russia, and the United States, defending their homelands. Jewish patriots founded the first and still

[9] Keller (2007), who builds on MacIntyre's analysis of patriotism as part of a family of "loyalty" virtues, makes this same mistake, when he essentially defines patriotism as a form of self-delusion or "bad faith" regarding one's country's good and bad qualities. By defining patriotism as an intellectual disability, Keller sets up a straw man, easily dismissed as a vice. But such trumped-up charges can be made against any of our particular moral commitments. The fact that some parents foolishly see only the good in their children, or even the fact (if it is a fact) that most parents are more inclined to see the good rather than the bad in them, hardly argues against parental commitment—given the importance of such commitment to human well-being. Similar considerations speak for a committed patriotism, because such commitment is the foundation for successful communities and societies. Of course, real patriots must be thinking patriots, just as good parents are thinking parents. Any virtue devolves into vice, when stripped of all reason.

the only real democracy in the Middle East. Patriots in Poland and Czechoslovakia eventually threw off communist tyranny.

Nussbaum's argument is marred by its simplified, inaccurate portrayal of patriotism. For example, she speaks of "the comfort of patriotism and its easy sentiments" (p. 7). But neither the American soldiers serving in Iraq, nor the American soldiers who have refused to serve there or who have chosen to speak out against the Iraq war once they return, are living lives of comfort and ease. Nussbaum's words are a cheap dismissal of those who struggle to make sense of patriotism's demands, or to balance the tensions between the demands of patriotism and ethical universalism. For example, those Israelis who love their country yet agonize over what justice demands for the Palestinians. Patriotism can of course be taken up lazily or unthinkingly, as a mere rationalization for selfish behavior. But so can ethical universalism. Let's kill the kulacks to speed on the formation of a classless society—and while we're at it, let's steal their stuff.

I agree with Benjamin Barber (another commentator): the answers to toxic patriotism or toxic cosmopolitanism are good patriotism and good cosmopolitanism. "What we require are healthy, democratic forms of local community and civic patriotism rather than abstract universalism and the thin gruel of contract relations" (Barber 1996, p. 31). And crucially, it is healthy democratic communities that allow us to instantiate the demands of ethical universalism. Securing human rights and a fairer distribution of resources only occurs when the proper civic infrastructure is in place, including a large percentage of individuals who think of themselves as *citizens*, concerned for the common good. These are the people who push for a living wage, or more affordable housing for their fellow citizens. Moral progress depends on the opportunities grasped by an engaged patriotism (Nelson 2002).

Again, for me, the environmental perspective brings this home with particular clarity. Those of us living in Fort Collins are blessed with a large number of local parks and natural areas to explore. We have these because half a dozen times in the past 25 years, city and county residents voted to tax themselves in order to buy and preserve the land. This depended, in turn, on two main factors. First, 15 or 20 key activists put in the time and effort to write the proposals and mount campaigns to pass the taxes. They succeeded only because of their deep knowledge of the local political terrain. Second, a majority of residents felt enough patriotic concern to vote to tax themselves to protect the land and preserve it for future generations (campaign ads usually included pictures of cute children hiking). I'm grateful to the super-patriots with the vision and determination to push for such initiatives, and to the patriotic-enough general populace who voted for them. I'm determined to do my part to preserve this heritage for those to come

So I lobby politicians, write checks and opinion pieces in the local papers, analyze dry technical documents, and send in comments to the Forest Service and the Corps of Engineers. I find these political efforts meaningful, but often tiresome. If I didn't find them meaningful, I wouldn't push through the drudgery. But their meaning comes in large part from their particularity. I'm working with my neighbors to protect the Cache la Poudre River, the same river that I've birded so many mornings. The river where I've skipped stones with my sons, and shown them their first ospreys and kingfishers, foxes and crawdads. The river along which I proposed to my wife. If my neighbors and I can protect our river, and if we succeed in building a community along its banks that lives in harmony with its wildlife, and if in 30 years I can walk along it and point out ospreys and otters to my grandchildren, then I will be well contented. Creating such communities and societies, linking the past to the future, preserving what is most valuable in our past and providing opportunities for future generations, are the goals of environmentalism. They demand focus.

No, responds Martha Nussbaum. The boundaries of a nation (and, we may assume, a city or a watershed) are "morally arbitrary." "Why should we think of people from China as our fellows the minute they dwell in a certain place, namely the United States, but not when they dwell in a certain other place, namely China? What is it about the national boundary that magically converts people toward whom we are both incurious and indifferent into people to whom we have duties of mutual respect?" (p. 14). I answer, first, that we need not be incurious or (completely) indifferent toward Chinese or other foreigners. This is a straw man. "Exclusive concern for one's own country is not a necessary part of patriotism" (Nathanson 1989, p. 538). Second, Chinese people who legally immigrate to the United States become fellow citizens, for whom we have special responsibilities by virtue of that tie. Such responsibilities include ensuring they have equal protection under our laws, access to sufficient health care when they are sick, etc. In return, Chinese–Americans take on those same civic responsibilities. Third, such responsible citizenship does not appear "magically." It must be cultivated; its absence has deleterious consequences, for society as a whole but particularly for its less fortunate members; and for that reason, we should think twice before undermining it, even unintentionally. Fourth, dividing the world up into smaller units called nations is one way to facilitate real, effective citizenship in an immense world of 6.7 billion people. In the same way, knowledge and devotion to particular landscapes makes environmentalism possible. Fifth, the primary responsibility for improving the lives of a nation's people resides in the people themselves, and in their leaders. If improvement is necessary, they need to take up the task themselves, as part of self-government. If they fail to do so, if they are selfish rather than patriotic, the efforts of outside benefactors are unlikely to succeed.

Earlier in her career, Nussbaum was keen to remind us that self-government is both a right and an essential human capability (Nussbaum 1993). But how this capability might be realized without a greater focus on one's compatriots than on complete strangers, or a greater concern for my town than some Chinese town halfway across the world, remains something of a mystery. It is interesting that once Nussbaum started writing about issues of third-world development and international justice, her earlier appreciation for self-government seemed to go by the board. Perhaps with so much suffering and gross injustice to alleviate in the third-world, specifying a proper first-world citizenship might seem like a luxury; much less important than raising awareness and dollars for overseas aid. "Giving one's money is a major way in which, in the absence of a world state, individuals can promote the good of those who are distant from them," Nussbaum (1996) writes (p. 134). This may be true. Yet there is a world of difference between giving charity to strangers and working out a common destiny with my fellow citizens. It seems to me that charity and citizenship are both necessary parts of a true humanity. But if that is true, then the moral claims of the wretched of the earth cannot be allowed to *displace* the moral claims, rights, and obligations of citizenship. At most, they are a necessary *supplement* to them.[10]

The hard truths behind Nussbaum's jibe at "arbitrary" boundaries is that they do lead to differential concern and action on behalf of others, and they do perpetuate differential access to resources. They lock in place the failures of societies, and the sins of the fathers are visited on the sons (and even more, on their daughters). *But borders also help lock in societies' successes* (Walzer 1983). They help preserve genuine human diversity in a world that has been shrunk and homogenized by modern technology. "Arbitrary" or artificial boundaries may be out of favor with the global intelligentsia. Yet they make self-government possible (Miller 1995) and self-government, we believe, is a key human capability and a fundamental human right.

Nussbaum finds it easy to dismiss patriotism because she nowhere grapples with the phenomenon itself and its complex, mixed impacts on human behavior. That patriotic affiliations might do real good in the actual world, perhaps inspire people to transcend their selfishness and act for the good of others, perhaps give their lives meaning, is irrelevant, since we can imagine better, more universalistic ideals, which would guide people to do the right thing, unerringly and for the right (universalistic) reasons. As her essay shows, this type of thinking coheres well with a managerial outlook (Pinsky 1996). The ultimate goal seems to be to turn flesh and blood human beings into efficient moral calculating machines (or, failing this, to free up the resources they are squatting on for more efficient distribution by their moral superiors). But when we consider what helps real people act morally, we find that our particular loyalties are important. For example, psychologists studying the virtue of justice have found that while a commitment to universal "fairness" helps lead people beyond selfishness, more partial strengths such as "loyalty" and "citizenship" are equally important in motivating just behavior (Peterson and Seligman 2004, Chap. 16).

All this is not to argue *against* more wide-ranging moral concerns and commitments. We may support both universal human rights and increased development aid for the world's poor, while also recognizing the legitimacy—indeed, the necessity—of people's particular moral affiliations. Such patriotic connections further and partly constitute full human flourishing, which should be the overarching goal of ethics. For that reason, patriotism is a virtue.

[10] Interestingly, Nussbaum, Gomberg (1990, p. 148), McCabe (1997, p. 218), and others often couple their criticisms of patriotism with demands that wealthy Westerners give more charity to poor people in the developing world. I'm sympathetic to the latter point; some increase in development aid for the world's poor does seem morally demanded today. But I think this way of asking for it, by disparaging the rights and responsibilities of particular citizenship, disrespects the humanity of everyone involved. It turns wealthy Westerners into cash machines and paints poor southerners as incompetent children, with neither the ability nor the responsibility for governing themselves.

Patriotism is Necessary

Love of country is a human possibility, which can be part of a fulfilling life. This possibility can also be ignored or misused. One of the worst misuses involves substituting a love of the state for a love of our land and our fellow citizens. Just as pernicious, some balance love of their own lands and peoples with contempt for others. Partly for these reasons, American intellectuals often see patriotism as a refuge for the simple-minded. Sometimes, indeed, it is such a refuge. Yet the principled understanding and retelling of our history is anything but simple. An honest patriot must wrestle with those aspects of our history of which we are ashamed. A compassionate patriot will remember history's losers, its dispossessed and despised.

Like courage, prudence, or indeed any virtue, patriotism is liable to a skewed development and to various kinds of misuse. Yet properly developed it is part of a good human life. Put another way, "patriotism" is a necessary word, but one whose meaning we must retrieve, by doing a little digging. As Henry Thoreau asserts: "It would seem as if the very language of our parlors would lose all its nerve and degenerate into *parlaver* wholly, our lives pass at such remoteness from its symbols, and its metaphors and tropes are necessarily so far fetched" (Thoreau 1971, pp. 244–245). Such has been the fate of "patriotism," a word coined by people who lived closer to the land than Thoreau's contemporaries, or ourselves. It has become a meaningless abstraction for many of us, in part through our mistaking abstractions for our true country. I think we must retrieve the word by rethinking and reliving it. When we live closer to the land, strive to know it better and work to protect all its inhabitants, human and nonhuman—that is patriotism.

The dangers of patriotism are undeniable. Yet so are the dangers stemming from a lack of patriotism. People need to band together in difficult times. We need to work with others to preserve what is important to us. We draw strength and inspiration from one another; we enjoy being part of larger efforts to achieve some common good. This emotional component need not overrule reason and morality. Like our other emotions, it is not something to suppress or be embarrassed about, but rather part of being human. It can be put to better or worse uses. In particular, patriotism can be put to good *environmental* uses, to bridge the liberal/conservative divide and achieve environmental protection. After three and a half decades of largely futile efforts to build on our early successes, American environmentalists, at least, should be willing to consider new ideas that might reinvigorate our movement.

In any case, patriotism is not as easily avoided as many intellectuals seem to think. It is a commonplace to say that environmentalism takes the place of religion, for many secular environmentalists. In a similar way, I think environmentalism takes the place of conventional patriotism among many who would recoil from the term. For lots of us, environmental work provides camaraderie and a sense of connection to something more important than ourselves. It is our way of contributing to our local communities and helping to define the meaning of America. Why not recognize this and call it by its proper name?

One Patriot

In *Walden*, Henry David Thoreau penned some of our most memorable words against patriotism, including these, from the concluding chapter:

> Every man is the lord of a realm beside which the earthly empire of the Czar is but a petty state, a hummock left by the ice. Yet some can be patriotic who have no *self*-respect, and sacrifice the greater to the less. They love the soil which makes their graves, but have no sympathy with the spirit which may still animate their clay. Patriotism is a maggot in their heads. (Thoreau 1971, p. 321)

For Thoreau, the individual self is greater than any state, and should never be sacrificed to it. Self-development and self-exploration should be our goals. *Walden* is the record of such self-exploration, deliberately cultivated far from the distractions of politics and the enervation of daily social contact (Cafaro 2004).

The passage seems to set patriotism—a sentiment of love for one's country—in opposition to these goals. Patriotism, Thoreau tells us, eats away at our brains, like a maggot, slowly but persistently. *Walden* argues that we must think our way toward better lives. Patriotism destroys this ability to think.

Many similarly dismissive references to a militaristic, thoughtless patriotism can be found in Thoreau's writings. Yet when it came time to speak out against injustice, he found it necessary to use the rhetoric of a patriot. In his lecture "Slavery in Massachusetts," Thoreau spoke angrily of the use of his state's militia to return the fugitive Anthony Burns to Virginia and slavery:

> I have lived for the last month,—and I think that every man in Massachusetts capable of the sentiment of patriotism must have had a similar experience,—with the sense of having suffered a vast and indefinite loss. I did not know at first what ailed me. At last it occurred to me that what I had lost was a country. (Thoreau 1973, p. 106)

As latter generations have often been reminded, our institutions of government may become instruments of injustice. But it is not only the threat to our own lives or happiness that moves us to fight injustice. We act because this is our country. We are ashamed of its injustices as we would be of our own, and concerned for what sort of country we are leaving our children.

Previous generations of Americans fought to end slavery and imperialist aggression. Thoreau argues that we must similarly fight to end the war against nature going on in our midst, and redefine our nation to include *all* its many places and inhabitants. In a posthumously published essay, he describes the sugar maples set up on Concord commons as a "*perfectly living* institution" (emphasis in the original):

> They are worth all and more than they have cost,—though one of the selectmen, while setting them out, took the cold which occasioned his death,—if only because they have filled the open eyes of children with their rich color unstintedly so many Octobers.... No annual training or muster of soldiery, no

> celebration with its scarfs and banners, could import into the town a hundredth part of the annual splendor of our October. We have only to set the trees, or let them stand, and Nature will find the colored drapery,—flags of all her nations. (Thoreau 1980, pp. 160, 165)

Here is an institution Thoreau can wholeheartedly believe in (unlike Congress, or the US Constitution); a benison to all, including future generations. Here, in the person of the anonymous selectman planting trees for his community, is an unheroic, unobtrusive service that we need to recognize as patriotism. A patriotism that is fully grounded yet expansive, not drawing lines defensively and saying "us or them," but widening our typical circle of moral concern and inviting in nature. Here is a patriotism that is truly a virtue.

There is ample scope for exercising such patriotism in planting trees, working to create new parks, or teaching children the names of the trees towering above them and the flowers at their feet. Most important, perhaps, is learning the stories of the places we inhabit and meeting the many "original settlers" (other species) with whom we still share this country. Not all the lessons learned are pleasant, of course. Nature is not all sunshine and ripe huckleberries, and some of the original settlers are gone for good. Attempts to effect political change are complicated and frustrating. Still we must strive to know the land and create *living* institutions; institutions in the service of Life. The alternative to such patriotic efforts is the loss of our country.[11]

References

Appiah, K. (1996). Cosmopolitan patriots. In J. Cohen (Ed.), *For love of country: Debating the limits of patriotism*. Boston: Beacon Press.

Barber, B. (1996). Constitutional faith. In J. Cohen (Ed.), *For love of country: Debating the limits of patriotism*. Boston: Beacon Press.

Bliese, J. (1997). Traditionalist conservatism and environmental ethics. *Environmental Ethics, 19*, 135–151.

Brown, L., et al. (Eds.). (1993). *The new shorter oxford English dictionary*. Oxford: Oxford University Press.

Cafaro, P. (2004). *Thoreau's living ethics: Walden and the pursuit of virtue*. Athens: University of Georgia Press.

Cafaro, P. (2005). Thoreau, Leopold, and Carson: Toward an environmental virtue ethics. In R. Sandler & P. Cafaro (Eds.), *Environmental virtue ethics*. Lanham, MD: Rowman and Littlefield.

Dewey, J. (1948). *Reconstruction in philosophy*. Boston: Beacon Press.

Driver, J. (1996). The virtues and human nature. In R. Crisp (Ed.), *How should one live? Essays on the virtues* (pp. 111–129). Oxford: Oxford University Press.

Gomberg, P. (1990). Patriotism is like racism. *Ethics, 101*, 144–150.

[11] For a more detailed discussion of Thoreau and patriotism, see Cafaro (2004), pp. 198–204. For a full account of Thoreau's environmental ethics, see the chapter titled "Nature" in the same volume. Many thanks to Ron Sandler, Kris Cafaro and two anonymous reviewers for the *Journal of Agricultural and Environmental Ethics*, for detailed comments which greatly improved this paper. Thanks also to audiences at Colorado College, Colorado State University and the University of Tulsa, whose comments also helped focus and improve the final version.

Intergovernmental Panel on Climate Change (IPCC) (2007). Fourth assessment report: Climate change 2007: Synthesis report: Summary for policymakers. Accessed at www.ipcc.org.
Keller, S. (2007). *The limits of loyalty*. New York: Cambridge University Press.
Leopold, A. (1949). *A sand county almanac and sketches here and there*. New York: Oxford University Press.
Louv, R. (2005). *Last child in the woods: Saving our children from nature-deficit disorder*. Chapel Hill, NC: Algonquin Books.
MacIntyre, A. (1984). *"Is patriotism a virtue?" Lindley lecture*. Lawrence: Kansas University Press.
McCabe, D. (1997). Patriotic gore, again. *Southern Journal of Philosophy, 35*, 203–223.
McConnell, M. (1996). Don't neglect the little platoons. In J. Cohen (Ed.), *For love of country: Debating the limits of patriotism*. Boston: Beacon Press.
McKibben, B. (2007). *Deep economy: The wealth of communities and the durable future*. New York: Henry Holt.
Millennium Ecosystem Assessment. (2005). *Ecosystems and human well-being: Biodiversity synthesis*. Washington, DC: World Resources Institute.
Miller, D. (1995). *On nationality*. New York: Oxford University Press.
Miller, R. (1998). Cosmopolitan respect and patriotic concern. *Philosophy & Public Affairs, 27*, 202–224.
Nathanson, S. (1989). In defense of moderate patriotism. *Ethics, 99*, 535–552.
Nelson, R. (2002). Patriots for the American land. In R. Nelson, B. Lopez, & T. Williams (Eds.), *Patriotism and the American land* (pp. 1–22). Great Barrington, MA: Orion Society.
Nussbaum, M. (1993). Non-relative virtue: An Aristotelian approach. In M. Nussbaum & A. Sen (Eds.), *The quality of life*. Oxford: Oxford University Press.
Nussbaum, M. (1996). Patriotism and cosmopolitanism" and "Reply. In J. Cohen (Ed.), *For love of country: Debating the limits of patriotism*. Boston: Beacon Press.
Peterson, C., & Seligman, M. (2004). *Character strengths and virtues: A handbook and classification*. Oxford University Press: Oxford.
Pinsky, R. (1996). Eros against Esperanto. In J. Cohen (Ed.), *For love of country: Debating the limits of patriotism*. Boston: Beacon Press.
Putnam, R. (2001). *Bowling alone: The collapse and revival of American community*. New York: Simon & Schuster.
Rolston, H., III. (1988). *Environmental ethics: Duties to and values in the natural world*. Philadelphia: Temple University Press.
Rorty, R. (1998). American national pride: Whitman and Dewey. In R. Rorty, (Ed.), *Achieving our country*. Cambridge: Harvard University Press.
Runte, A. (1979). *National parks: The American experience*. Lincoln: University of Nebraska Press.
Sandler, R. (2007). *Character and environment: A virtue-oriented approach to environmental ethics*. New York: Columbia University Press.
Schmidtz, D. (1997). Self-interest: What's in it for me? In E. Frankel Paul, F. Miller, & J. Paul (Eds.), *Self-interest*. Cambridge: Cambridge University Press.
Shi, D. (1985). *The simple life*. New York: Oxford University Press.
Soukhanov, A. (Ed.). (1992). *The American heritage dictionary of the English language* (3rd ed.). Boston: Houghton Mifflin.
Swanton, C. (2003). *Virtue ethics: A pluralistic view*. Oxford: Oxford University Press.
Thoreau, H. (1971). *Walden*. Princeton: Princeton University Press.
Thoreau, H. (1973). *Reform papers*. Princeton: Princeton University Press.
Thoreau, H. (1980). *The natural history essays*. Salt Lake City: Peregrine Smith Press.
Walzer, M. (1983). *Spheres of justice: A defense of pluralism and equality*. (New York: Basic Books.
Walzer, M. (1996). Spheres of affection. In J. Cohen (Ed.), *For love of country: Debating the limits of patriotism*. Boston: Beacon Press.
Wensveen, L. V. (2001). Attunement: An ecological spin on the virtue of temperance. *Philosophy in the Contemporary World, 8*(2), 67–78.

GPSR Compliance

The European Union's (EU) General Product Safety Regulation (GPSR) is a set of rules that requires consumer products to be safe and our obligations to ensure this.

If you have any concerns about our products, you can contact us on ProductSafety@springernature.com

In case Publisher is established outside the EU, the EU authorized representative is:

Springer Nature Customer Service Center GmbH
Europaplatz 3
69115 Heidelberg, Germany

Batch number: 10823959

Printed by Printforce, the Netherlands